Japan's Nuclear Crisis

Japan's Nuclear Crisis

The Routes to Responsibility

Susan Carpenter

palgrave
macmillan

First published 2012 by
PALGRAVE MACMILLAN

Palgrave Macmillan in the UK is an imprint of Macmillan Publishers Limited, registered in England, company number 785998, of Houndmills, Basingstoke, Hampshire RG21 6XS.

Palgrave Macmillan in the US is a division of St Martin's Press LLC, 175 Fifth Avenue, New York, NY 10010.

Palgrave Macmillan is the global academic imprint of the above companies and has companies and representatives throughout the world.

Palgrave® and Macmillan® are registered trademarks in the United States, the United Kingdom, Europe and other countries.

ISBN 978–0–230–35492–0

This book is printed on paper suitable for recycling and made from fully managed and sustained forest sources. Logging, pulping and manufacturing processes are expected to conform to the environmental regulations of the country of origin.

A catalogue record for this book is available from the British Library.

A catalog record for this book is available from the Library of Congress.

10 9 8 7 6 5 4 3 2 1
21 20 19 18 17 16 15 14 13 12

Transferred to Digital Printing in 2012

Contents

Timeline

At 14.46 PM local time Tohoku, Honshu's northeast region, was hit by an earthquake with a magnitude of 9.0	11th March 2011	Earthquake hits Japan
On the evening of 11 March, the government declared a state of emergency at the Fukushima Daiichi Nuclear Power Plant. TEPCO workers are unable to reconnect electricity to the cooling system.	11th March 2011	State of emergency at Fukushima
In the early evening of 12 March a hydrogen explosion occurred in reactor No. 3 and blew the roof off the containment vessel housing reactor No.1.	12th March 2011	Hydrogen explosions at reactor 3
On 15 March two more explosions rocked the plant and a fire broke out in No. 4 reactor.	15th March 2011	Two more explosions at plant
On 17 March, winds carrying mildly radioactive smoke and dust raised radioactivity 30 times above normal levels in Tokyo.	17th March 2011	Radioactivity detected in Tokyo
On 18 March damage to No. 3 reactor was discovered.	18th March 2011	No. 3 reactor core damaged

Date	Event	Summary
10th April 2011	On 10 April a worker from No. 2 reactor was admitted to hospital after he complained of illness.	Worker from Fukushima in hospital
12th April 2011	On 12 April NISA raised the Fukushima crisis to a level 7, the same level as Chernobyl.	NISA raise crisis level
14th April 2011	On 14 April TEPCO announced that radiation levels had climbed in reactor No. 4's spent fuel rods.	TEPCO announce radiation increase
15th April 2011	On 15 April government announced that levels of Iodide-131 found in sea water near the plant had spiked from 1,100 times the previous day to 6,500 times the legal limit.	Harmful Iodide-131 levels climb
30th April 2011	On 30 April the Ministry of Health, Labor and Welfare (HLW) reported that tiny amounts of Iodide-131 had been found in 7 samples of breast milk of women living in Fukushima, Ibaraki and Chiba	Iodide-131 found in breast milk
2nd May 2011	TEPCO reports radiation in the seabed near Daiichi rises to 100–1,000 times above the normal levels	Radiation in seabed near Daiichi increases

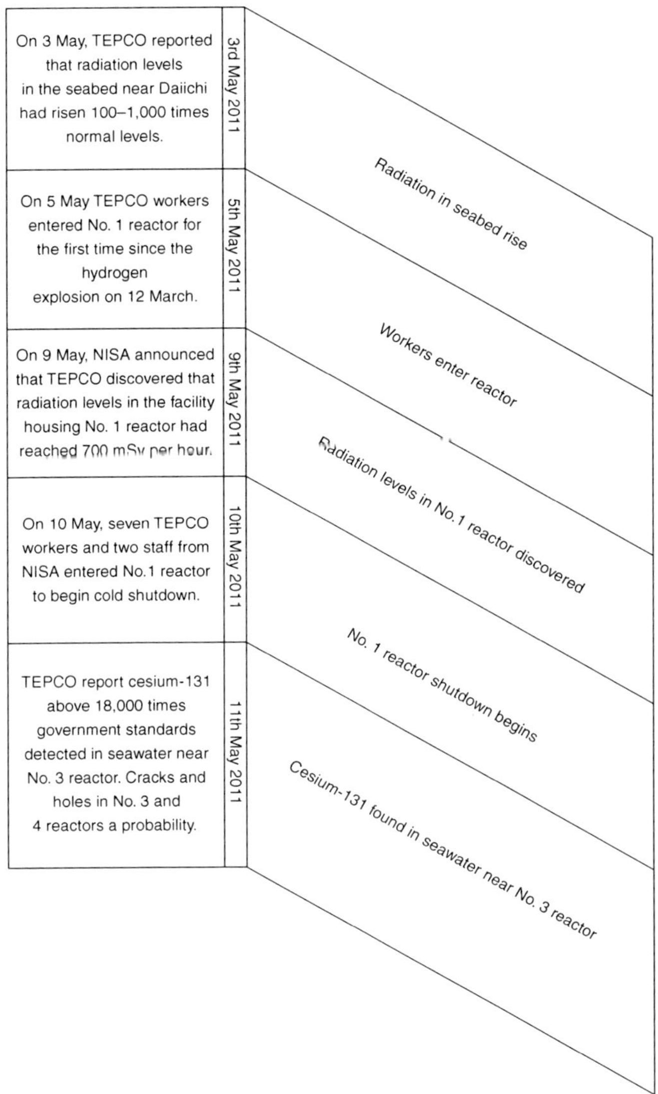
On 3 May, TEPCO reported that radiation levels in the seabed near Daiichi had risen 100–1,000 times normal levels.
3rd May 2011
Radiation in seabed rise
On 5 May TEPCO workers entered No. 1 reactor for the first time since the hydrogen explosion on 12 March.
5th May 2011
Workers enter reactor
On 9 May, NISA announced that TEPCO discovered that radiation levels in the facility housing No. 1 reactor had reached 700 mSv per hour.
9th May 2011
Radiation levels in No.1 reactor discovered
On 10 May, seven TEPCO workers and two staff from NISA entered No.1 reactor to begin cold shutdown.
10th May 2011
No. 1 reactor shutdown begins
TEPCO report cesium-131 above 18,000 times government standards detected in seawater near No. 3 reactor. Cracks and holes in No. 3 and 4 reactors a probability.
11th May 2011
Cesium-131 found in seawater near No. 3 reactor

Date	Event	Summary
12th May 2011	NISA reports meltdown of fuel rods in No.1 reactor. Holes and cracks in all reactors delay decontaminations and shutdown.	Reports of meltdown of fuel rods in No.1 reactor
13th May 2011	NISA reports meltdowns in No. 2 and No. 3 reactors. No. 3 reactor fuel rods contain 6% MOX fuel.	Reports of meltdowns in No. 2 and 3 reactors
20th May 2011	40 tons of highly radioactive water from No. 3 reactor leaked into ocean.	Water from No. 3 reactor leaked into ocean
26th May 2011	TEPCO reports that the cooling system had been damaged by the earthquake and not the tsunami.	Earthquake cause of cooling system damage
30th May 2011	TEPCO announces that due to holes in the container vessels caused by the meltdowns, the plant will not be brought under control by the year end.	Power plant unlikely to be under control by end of the year

Date	Event	Summary
3rd June 2011	Radiation levels outside No. 1 reactor registers 4,000 mSv per hour, the highest reading since 12 March.	Radiation outside No.1 reactor rises
6th June 2011	Radiation level from Daiichi was 370,000 tbq, during week of 12 March suggesting meltdowns occurred much quicker than originally thought.	NISA reports meltdowns occurred much quicker than expected
8th June 2011	The government announces radiation found in soil samples taken in March–April 62km from Daiichi.	Radiation found in soil 62km from Daiichi
15th June 2011	TEPCO reports accumulation of 2,000 cubic metres of highly radioactive sludge by year end.	TEPCO estimates amount of radioactive sludge by year end
17th June 2011	TEPCO announces the start of the release of 100,000 metric tonnes of radioactive contaminated water into the ocean.	Contaminated water released into the ocean

Timeline courtesy of Lawrence Tait, The University of Edinburgh School of Engineering.

Annual Average Exchange Rates

Japanese Yen per United States Dollar [JPY/USD]

2000: 106.46

2001: 122.37

2002: 124:79

2003: 115.82

2004: 107.49

2005: 110.85

2006: 116.12

2007: 117.27

2008: 102.80

2009: 93.50

2010: 87.10

Acronyms and Abbreviations

AIST	National Institute of Advanced Industrial Science and Technology
ANRE	Agency of Natural Resources and Energy
BOJ	Bank of Japan
BP	Basic Pension
CHUDEN	Chubu Electric Power Co.
CPI	Consumer Price Index
DBJ	Development Bank of Japan
DPJ	Democratic Party of Japan
EC	European Commission
EPDC	Electric Power Development Co.
FAZ	Foreign Access Zone
FBR	Fast Breeder Reactor
FDI	Foreign Direct Investment
FEPC	Federation of Electric Power Companies
FILP	Fiscal Investment and Loan Program
FRI	Fujitsu Research Institute
GDP	Gross Domestic Product
GHLC	Government Housing Loan Co.
HEPCO	Hokkaido Electric Power Co.
IAEA	International Atomic Energy Agency
IAI	Independent Administrative Institution
IBJ	Industrial Bank of Japan
IMF	International Monetary Fund
JAEA	Japan Atomic Energy Agency
JAERI	Japan Atomic Energy Research Institute
JAPCO	Japan Atomic Power Co.
JBIC	Japan Bank of International Cooperation

JDP	Japan Development Bank
JETRO	Japan External Trade Organization
JFTC	Japan Fair Trade Commission
JGB	Japan Government Bonds
JH	Japan Highway Co.
JHF	Japan Housing Finance Agency
JNC	Japan Nuclear Cycle Development Institute
JNEPA	Japan Newspaper Editors and Publishers Association
JNES	Japan Nuclear Safety Organization
JNFL	Japan Nuclear Fuel Cycle Co. Ltd.
JNOC	Japan National Oil Co.
JOGMEC	Japan Oil, Gas and Metals National Co.
JP	Japan Post
JPS	Japan Pension Service
JPSS	Japan Postal Savings System
Keidanren	Japan Business Association
KEPCO	Kansai Electric Power Co.
KYUDEN	Kyushu Electric Power Co.
LDP	Liberal Democratic Party of Japan
LTCB	Long-Term Credit Bank
MAC	Ministry of Agriculture and Commerce
MCI	Ministry of Commerce and Industry
METI	Ministry of Economy, Trade and Industry
MIC	Ministry of Internal Affairs and Communications
MLIT	Ministry of Land, Infrastructure and Transportation
MOC	Ministry of Construction
MOD	Ministry of Defense
MOX	Mixed Oxide Fuel
MEXT	Ministry of Education, Culture, Sports, Science and Technology
MHLW	Ministry of Health, Labor and Welfare
MM	Ministry of Munitions

MOF	Ministry of Finance
MOFA	Ministry of Foreign Affairs
MOX	Mixed Oxide Fuel
mSv	millisieverts
NISA	Nuclear Industrial Safety Agency
NPL	Non-performing loans
NSC	Nuclear Safety Commission
NSK	Nihon Shimbun Kyokai
NTT	Nippon Telephone and Telegraph
NUMO	Nuclear High-Level Waste Management Organization
OECD	Organisation for Economic Co-operation and Development
QE	Quantitative Easing
SCAP	Supreme Commander of the Allied Powers
SDF	Self Defence Force
SIA	Social Insurance Agency
TEPCO	Tokyo Electric Power Co.
TIRB	Temporary Industrial Rationalization Bureau
TODEN	Tohoku Electric Co.
TSE	Tokyo Stock Exchange
UDC	Urban Development Co.
UNCTAD	United Nations Conference on Trade and Development
URO	Urban Renaissance Organization
USTR	United States Trade Representative
WTO	World Trade Organization
YODEN	Shikoku Electric Power Co.

Introduction

On 11 March 2011, Tohoku, Honshu's northeast region, was hit by an earthquake with a magnitude of 9.0. The quake's epicenter was 70 km east of the Oshiku Peninsula and 250 km northeast of Tokyo. It was the most powerful on record in Japan and the fifth most powerful in the world. The Tokai and Kanto regions were also badly shaken. The earthquake caused severe structural damage to homes, shops, roads, bridges, railway lines, factories, schools and other infrastructure. A thirty-foot-high tsunami followed so swiftly that local residents whose homes and businesses were on the coast could not escape to higher ground. The tsunami destroyed entire villages along the coast and, according to the National Police report issued on 1 June, 15,281 people lost their lives, 8,492 people were missing, 5,363 were injured, 150,000 were homeless[1] and 3,970 roads and 71 bridges were destroyed. The government estimated that the damage caused by the disaster was ¥17 billion ($211 billion).

On the evening of 11 March, the government declared a state of emergency at the Fukushima Daiichi Nuclear Power Plant. The power station is located on the coast of Fukushima Prefecture, about 45 km south of Sendai City, near the quake's epicenter. The tsunami tidal wave flooded the diesel generators, causing them to fail. Since there was no back-up power system installed, the reactors' cooling system shut down and temperatures in four of the reactors began to rise. Tokyo Electric Power Company (TEPCO), the plant's operator, struggled to reconnect the electric power to the pumping system to cool down the reactors, which contained several hundred thousand extremely radioactive fuel rods. In August 2010, TEPCO had started to load No. 3 reactor with 6 percent Mixed Uranium-Plutonium Oxide fuel (MOX), which is a highly toxic and combustible mixture of uranium and plutonium oxides.

TEPCO workers, unable to reconnect the electricity to the cooling system, resorted initially to pumping water from fire trucks into the reactor containers. The following day the radiation level in No. 1 reactor was reported to be rising steadily and Japan's Nuclear Industrial and Safety Agency (NISA)[2] announced that radiation levels at the plant's main gate were eight times above the normal level. By the afternoon it was apparent that small amounts of radioactive cesium had escaped from the plant. The water levels in No 1. reactor's core fell and exposed the four fuel rods. There were three hydrogen explosions and the reactors leaked radiation into the soil within a 60 km radius of the plant. A brief chronology of the events from March through August will reveal that TEPCO was either unaware of the seriousness of the event and the approaching disaster or reluctant to release pertinent information.

12 March: The fuel rods melted and fragments of the damaged rods fell to the bottom of the core container, which created holes in the pressure container. Early in the evening a hydrogen explosion occurred in No. 1 reactor, blowing the roof off the containment vessel and injuring four people and spewing radioactive steam into the atmosphere. 230,000 residents living in the vicinity of the plant were evacuated.

14 March: A hydrogen explosion rocked No. 3 reactor.

15 March: Another explosion occurred in No. 2 reactor and a fire broke out in No. 4 reactor. Radiation levels around the facility spiked to the dangerous level of 400 millisieverts (mSv).[3] It was evident that there were meltdowns in several of the four reactors. The government advised residents living within a 20 km radius of the plant to stay indoors.

17 March: Gregory Jacszko, the chairman of the United States Nuclear Regulatory Commission, denounced the 20 km radius and strongly recommended that a 50 km exclusion zone should be established. Foreign governments also urged nationals to either leave the region or return to their countries.

Winds carrying mildly radioactive smoke and dust raised radioactivity thirty times above normal levels in Tokyo. Japanese families with children, who were considered most vulnerable to the radioactivity, began a mass migration to regions south of Tokyo. Foreign governments with embassies in Tokyo, including the US and Britain, moved staff to offices in Osaka and foreign companies with corporate offices in Tokyo advised nationals and staff to leave Tokyo until the crisis was under control and there was no longer a risk of contamination from radioactivity. The US State Department along with other foreign governments gave authorization to US embassy family members to leave on a voluntary basis. Foreign air carriers discontinued flights over Japan and, although some carriers returned to normal service within three weeks, others were reluctant to return to normal service until mid-April. The US finally approved the resumption of flights to Sendai Airport on 21 July.

18 March: Damage to No. 3 reactor core was discovered and the Japan Nuclear Energy Safety Organization (JNES)[4] raised the level of the crisis from four to five on the international scale of the seriousness of incidents. The highest is seven. The event at Three Mile Island was regarded as a five and at Chernobyl as a seven. Residents who were living within a 20 km radius of Daiichi were evacuated.

19 March: Abnormal levels of iodide-131 radioactivity were detected in milk and spinach produced near Daiichi. Iodide-131 was also found in Tokyo tap water.[5]

21 March: The government ordered the halt of shipments of certain produce from four prefectures after higher than normal levels of radioactivity were detected in the produce. TEPCO, which is the fourth largest electric power company in the world, the largest power company in Asia, the largest supplier of electricity in Japan and the major supplier of electric power to Tohoku, Tokai and Kanto, was forced to cut electric power by 25 percent. Since other power plants were also damaged the government ordered a rolling blackout in the Tohoku–Kanto regions throughout the summer months. Manufacturers with large operations in Tohoku, Tokai and Kanto regions who were trying to restart operations that had been halted by the earthquake were forced to shut down plants temporarily.

22 March: When the government warned parents in Tokyo that tap water was unsafe for infants and to substitute bottled water, supplies disappeared from market shelves and convenience stores in Tokyo and the surrounding prefectures. The government lifted the warning a week later. Nevertheless, residents remained vigilant and continued to hoard bottled water. Governments in Fukushima, Ibaraki, Chiba, Saitama and Tochigi prefectures also detected higher than normal levels of iodide-131 in drinking water and issued warnings.

24 March: Three workers who stepped into a puddle of highly radioactive water while trying to connect electric cables to cooling pumps received high doses of radiation. Two were hospitalized after complaining of illness. They were discharged on 27 March with no signs of serious illness.

29 March: Radioactive water leaked from an eight-inch crack in one of the containment vessels into the Pacific Ocean before the crack was plugged on 6 April but small leakages of radioactivity continued to frustrate TEPCO's attempts to bring the plant under control.

5 April: Workers pumped 3 million gallons of radioactive water into the ocean.

7 April: TEPCO began injecting nitrogen into No. 1 reactor to prevent another hydrogen explosion.

8 April: An aftershock with a magnitude of 7.4 caused another large-scale blackout creating further disruption to manufacturers who were trying to restarting operations.

10 April: A worker in No. 2 reactor was admitted to a hospital after he complained of illness but he was later released having suffered no apparent ill effects. As water was pumped into the crippled reactors, the amount of radioactive water increased. TEPCO released 11,500 tons of radioactive water into the Pacific on 4 April in order to make space for the contaminated water in No. 2 reactor, which was still leaking radioactive water into the ground soil around the plant. Within a two-week period TEPCO had released 60,000 tons of contaminated water into the Pacific.

11 April: On the one-month anniversary of the 9.0 quake, an after-shock with a magnitude of 7.1 hit the Fukushima coast. On 12 April a 7.0 aftershock shook the same region.

12 April: NISA raised the Fukushima crisis to a level seven, the same level as Chernobyl, because of the impact of the cumulative effects from the continuing leaks of radioactive water and the widespread health and environmental effects. An official at TEPCO suggested that since the leaks had not ceased, the situation could be worse than Chernobyl. Another earthquake of magnitude 6.0 struck the same day. After each aftershock workers who were trying to stabilize the reactors had to evacuate for fear of more explosions and spiking emissions of radiation.

To underscore the seriousness of the situation, on 14 April TEPCO announced that radiation levels had climbed in No. 4 reactor's spent fuel rods and that temperatures had risen to 94°C. When No. 4 caught fire on 15 March the temperature was 84°C.

15 April: The government announced that levels of iodide-131 found in seawater near the plant had spiked from 1,100 times the previous day to 6,500 times the legal limit. The levels of cesium-134 and cesium-137 had risen nearly four times, suggesting that new leaks had developed.

The following day TEPCO announced that it aimed to bring down radiation levels within three months and put the plant into cold shut-down within nine months to end the crisis. However, intermittent rises in temperatures in the reactors, leading to more leaks, were posing ongoing problems and hindered the attempts of exhausted plant workers to stabilize the reactors. Gregory Jacszko pronounced the situation at the plant as 'static' not 'stable'.

30 April: The Ministry of Health, Labor and Welfare (MHLW) reported that tiny amounts of 131-iodide had been found in seven samples of

breast milk of women living in Fukushima, Ibaraki and Chiba prefectures. There was no designated safety standard for radioactive substances in breast milk.

On the same day, University of Tokyo Professor Toshiso Kosako, an expert on radiation safety and who had been appointed by the Prime Minister in March as one of six senior science advisors on radiation safety limits for schools located in the vicinity of Daiichi, resigned. Kosako told reporters that the government was ignoring the laws regarding managing nuclear disasters and that improvised measures were serving to prolong the crisis. He condemned the government for a lack of transparency regarding the dangerous levels of radiation.

2 May: The American non-profit organization Physicians for Social Responsibility (PSR) denounced the Japanese government's safety standards of 20 mSv for radiation levels at elementary and junior high schools. Pronouncing the standards 'unconscionable' the organization emphasized that children and fetuses were more susceptible to radiation than adults. The PSR emphasized that radiation of 20 mSv over a one-year period would expose them to a 1 in 200 chance of developing cancer. Exposure for two years would increase the risk to 1 in 100.

3 May: TEPCO reported that radiation levels in the seabed near Daiichi had risen 100–1,000 times normal levels.

4 May: As TEPCO continued to struggle to bring Daiichi under control, Jacszko remained cautious about the stability of Daiichi, telling the United States House Energy and Commerce Committee: 'While we have not seen or predicted any significant challenges to safety at the site, we have only seen incremental improvements towards stabilizing the reactors and spent fuel pools.'

5 May: TEPCO workers entered the No. 1 reactor for the first time since the hydrogen explosion occurred outside the reactor's container on 12 March. The radiation level was 19 mSv per hour. The workers connected eight pipes to a ventilating machine situated at the turbine building next to the reactor in order to reduce the high level of radiation contamination. The objective was to prevent workers from inhaling radioactive particles and suffering internal exposure when they entered the building to install heat-exchange equipment.

On the same day, media announced that Daiichi workers were very concerned about the dangers of high radiation exposure and the poor working conditions in the plant. Workers also complained that many areas were filled with highly radioactive rubble that had been strewn around the compound by the hydrogen blast. Workers began refusing to continue to work at the plant and TEPCO was having difficulties sourcing staff to enter Daiichi.

9 May: NISA announced that TEPCO had discovered that radiation levels in the facility housing No. 1 reactor had reached 700 mSv per hour, making the working environment difficult, and that radiation shielding was needed in order to bring the crisis to an end. Safety regulations for workers in nuclear installations prohibit the accumulation of more than 20 mSv annually or 100 mSv over a five-year period. However, the government revised the standard to 250 mSv to allow workers at Daiichi to bring the plant under control but workers were not allowed to spend more than twenty-minute intervals in the areas where the levels of radiation were highest. The government placed a ban on the sale of popular Fukushima bamboo shoots and wild vegetables grown in areas near Daiichi because of the unsafe levels of cesium.

10 May: Seven TEPCO workers and two staff from NISA entered No. 1 reactor to begin the cold shutdown.

11 May: There was a new leak of highly radioactive water into the ocean near Daiichi. TEPCO said that cesium-131 at above 18,000 times government standards was found in seawater near No. 3 reactor. TEPCO found the leak coming from a pit connected to a trench next to the reactor and plugged it with concrete. However, iodide-131 at 85,000 times the legal limit and cesium-134 at levels 620,000 times the legal limit were detected inside the pit. Cracks and holes in both No. 3 and No. 4 reactors were probable.

12 May: NISA officials reported that the repaired monitoring equipment showed that the water level in No. 1 reactor's pressure vessel was much lower than had been anticipated and that there had been a meltdown of the fuel rods which fell to the bottom of the container. It was probable that the meltdown had occurred sixteen hours after the earthquake struck.

The low water level indicated that there could be holes or cracks in the pressure vessel through which radioactive water was leaking and that the cracks were caused by the earthquake and not by the tsunami. Also the water level in the containment vessel that workers had been spraying to cool down the pressure vessel from the outside was lower than expected, indicating that there could be holes in that vessel as well and in the components connected to it. Holes or cracks in the vessels of all of the reactors would make repair work more difficult and delay the cold shutdown.

13 May: NISA reported that the monitoring equipment had incorrectly gauged the low water levels at both No. 2 and No. 3 reactors and that

there had been meltdowns in those reactors and pieces of the fuel rods had sunk to the bottom of the pressure vessels. The fuel rods in No. 3 reactor contained 6 percent MOX fuel. No. 2 reactor was particularly volatile. The many holes and cracks that had been found in all of the pressure vessels would make repairs extremely difficult because of high radiation levels in the containers. Furthermore, the accumulated seawater in the vessels was leaking through the holes and cracks, and increasing the amounts of radioactive water trapped in the pits below the vessels.

20 May: TEPCO announced that during May, 50 tons of highly radioactive water had leaked into the ocean from No. 3 reactor. The announcement coincided with the visit to Fukushima City by Chinese and South Korean leaders who expressed deep concerns about the delay of the release of information to their countries regarding the conditions at Daiichi.

26 May: TEPCO announced that the cooling system had been damaged by the earthquake and not by the tsunami, which had been presumed initially.

30 May: A typhoon hit Fukushima and heavy rains forced TEPCO to halt some of the outdoor work on Daiichi. The utility announced that due to the holes and cracks in the container vessels caused by the meltdown of the fuel rods in No. 1, 2 and 3 reactors it would be impossible to bring the plant under control by the end of the year.

3 June: Radiation levels outside No. 1 reactor registered 4,000 mSv per hour, the highest reading since 12 March. Steam was also spewing from a pipe under the reactor's floor where highly radioactive contaminated water had accumulated. Workers in the reactor at the time were exposed to 4 mSv of radiation.

6 June: NISA announced that the level of radiation released from Daiichi into the atmosphere during the week of 11 March was twice the amount originally estimated, which was 370,000 terabecquerels (Tbq)[6] suggesting that the meltdown of the fuel rods occurred much sooner than had been assumed at the time.

8 June: The government announced that radiation was detected in soil samples taken in March and April 62 km from Daiichi.

15 June: TEPCO confessed that the decontamination of highly radioactive water which had accumulated when seawater had been pumped into the reactors to cool them would result in about 2,000 cubic meters of highly radioactive sludge by year-end. The disposal site for the sludge was yet to be determined.

17 June: TEPCO announced that it would begin releasing 100,000 metric tonnes of radioactive contaminated water into the sea using French and American technologies.

26 June: TEPCO admitted it had not reported to NISA on 14 March data that indicated that hydrogen was accumulating in No. 3 reactor because the fuel rods were exposed to air and that there was the possibility of a hydrogen explosion.

30 June: The government identified 113 households in Date, Fukushima, a town that is located 80 km from Daiichi, as radioactive hotspots and recommended that the residents evacuate.

Radioactive sludge began to accumulate as TEPCO pumped the contaminated water from the reactor vessels throughout the summer. By August, over 1,500 metric tonnes of highly radioactive sludge was discovered in five prefectures. The sludge was too contaminated to be buried and local authorities were struggling to find appropriate storage sites.

1 August: TEPCO announced that the highest level of radiation since 11 March was detected at the piping that connected the main exhaust tower between No. 1 and 2 reactors. The level of 10,000 mSv per hour indicated the possibility of high levels of radiation around the plant and that workers were being exposed to much higher doses of radiation outside the vessels than had been considered initially.

8 August: The Japanese media reported that a senior researcher at the former Japan Atomic Energy Institute (JAERI), which developed MOX, considered that there had been a second meltdown of the MOX fuel in No. 3 reactor on 21 March, triggering the release of large amounts of radioactivity into the atmosphere.

Consequences of the nuclear crisis: agriculture, fisheries

The businesses of Fukushima farmers and fishermen suffered huge losses from the ongoing crisis. Fukushima is the fourth largest producer of vegetables and rice in Japan. In 2009, the industry accounted for 2 percent of Japan's GDP.[7] The fishery industry is also a primary industry. Directly following the explosions and leaks, produce, milk and fish from Fukushima were removed from Tokyo supermarkets after the discovery that they contained levels of iodide-131 exceeding government standards. In addition, the government imposed a ban on rice planting in areas located beyond the 30 km exclusion zone in Fukushima. It was the first time that farmers had been issued orders to curtail planting. Prime Minister Naoto Kan also banned mushrooms grown in eastern Fukushima and ordered the slaughter of livestock raised in the 20 km evacuation zone.

The radioactive contamination affected agriculture in other prefectures as well. A few days after radiation was discovered in Fukushima produce, higher than normal levels of iodide-131 were discovered in vegetables produced in Miyagi, Chiba, Ibaraki, Iwate, Gunma, Yamagata and Tokyo and were removed from market shelves.

By 9 June, according to the MHLW, the list of food products that registered radiation levels above government safety limits increased to 347 from eight prefectures. Products included spinach, mushrooms, bamboo shoots, tea, plums and fish. Japan's green tea industry was also adversely affected by the continued radioactive fallout. Radiation was detected in tea leaves grown in Shizuoka Prefecture which is 400 km from Fukushima and later in Tokyo tea leaves, impacting on the industry's domestic and foreign sales. The Japan Tea Growers Association protested vigorously about the government's regulations to counteract the bad publicity.

Apple growers in Aomori Prefecture saw their exports to Taiwan in May decrease to less than one ton. Aomori Prefecture, which is located in Tohoku on the Sea of Japan, produces 90 percent of apples in Japan. The apples, which are reputed to be the world's most delicious variety, comprise 70 percent of Japan's total fruit exports. Aomori apples are highly prized by the Taiwanese who are willing to pay top prices for them. The farmers were concerned that the continuation of the nuclear crisis together with the strong yen would impact on exports and force an increase of sales in the domestic market and thus a decline in prices.

The cattle farming industry was particularly affected by radioactive fallout from Daiichi. In July, the Tokyo metropolitan government announced that high levels of radiation had been detected in the meat of ten cattle fed contaminated straw feed and shipped from Fukushima Prefecture to a Tokyo meat packing plant. The meat had been processed and distributed to markets and restaurants in six prefectures, including Tokyo and Osaka. Aeon, Japan's second largest retailer, also announced that it had distributed 703 pounds of cesium-contaminated beef from Fukushima. On 18 July, the government suspended the shipment of cattle from Fukushima. A few days later, the government reported that 1,235 cattle at farms located throughout Japan had been fed contaminated feed grown in Fukushima and seven other prefectures. The feed was not on the list of contaminated products because it was not considered a food source. The problem escalated the following day when the government announced that 648 cattle suspected of being contaminated with radioactive cesium had been shipped to 43 prefectures. The cattle showed levels of 4,350 Tbq per kilogram. The safety limit is 500 Tbq. The beef scare enveloped more prefectures when high levels of cesium were found in cattle that had been fed on contaminated feed grown on farms located over 150 km from Fukushima.

However, the losses to farmers and fishermen did not stop at the domestic market. When traces of iodide-131 were first discovered, China, a key importer of Japanese foods, banned imports of foods from Fukushima but, as the crisis continued into April, China imposed a new ban of imported foods from Tokyo, Niigata, Nagano, Saitama, Tochigi and Yamanashi prefectures as well. Sales of spinach in China fell 50 percent in April. By May, China had introduced new regulations that required that food imports from other regions of Japan had certificates guaranteeing that the foods were free of radiation. The Japanese embassy in Beijing reported that the regulations were slowing the entry of imports.

South Korea also required certificates issued by the Japanese government and stricter safety checks. The Philippines in May banned imports of all produce from Japan. Consequently, in May, food imports fell 74 percent to 75 tons from 293 tons in April. Thailand was also wary of fresh food imports from Japan. The Food and Drug Administration required importers to certify the safety of vegetables, fruits, meats and marine products from the twelve prefectures. Although India's Health Ministry announced a three-month ban on imports, it claimed that it was making only a recommendation but requested that the Japanese government certify that the foods were free of radioactive substances.

Despite the promises from TEPCO and the government to compensate farmers for their losses, estimated to be well over $1 billion, Fukushima farmers and fishermen were adamant that their industries and livelihoods had been ruined and that businesses would never be able to recover because Japanese and foreign consumers would no longer trust their brands. Farmers based in the other prefectures were also very concerned that the ongoing crisis at Fukushima would make a lasting impression on consumers and, even if radioactive levels in foods were well below the standard set by government, confidence in Japanese food exports would quickly deteriorate. Businesses were particularly worried about the ban on exports, which could continue for months if the crisis was not brought under control. Also, the government's ad hoc approach to testing the levels of radiation in food products made Fukushima residents wary of purchasing any fresh farm products grown not only in Fukushima but also in the other six prefectures.

The Tohoku region produces about 20 percent of domestic wheat, dairy, meat, rice and vegetables. Due to domestic fears of contamination, Japan expected to import twice the amount of foods and that the imports from China would increase dramatically.

Consequences of the nuclear crisis: Fukushima residents

One month after they had been evacuated, the government declared that the 20 km evacuation zone would be officially off-limits and that under the law on nuclear disaster counter-measures, the government could order heads of cities, towns and villages to prohibit people from entering. The local authorities could also order people to leave under the threat of punishment. The no-go zone was extended to 30 km by May. However, the United States recommended that US citizens stay at least 80 km from the plant.

The majority of the 90,000 evacuees may never be able to return to their homes. On 15 April the government requested that TEPCO pay preliminary compensation of ¥170,000 ($11,800) to each of the 50,000 households for immediate living expenses. The evacuees protested that the amount was completely inadequate. Although Japanese law exempts nuclear power plants from paying compensation if the damage to a reactor results from 'a grave natural disaster of an exceptional character', the government announced on 2 May that there would be no cap on compensation claims because such a disaster had occurred in the past and because the risk of another occurrence had been recognized. In March, JP Morgan calculated that due to compensation payments, TEPCO's losses in FY 2010 could total ¥2 trillion ($25 billion). Bank of America-Merrill Lynch estimated that if the crisis continued losses could be $130 billion, which was the more accurate estimate. In addition, the number of temporary homes that the government promised to build for the evacuees by the end of August fell short of the initial target.

The ongoing crisis created an atmosphere of uneasiness among the local population who feared the long-term effects of exposure to the radiation leaking from the crippled plant. The MHLW reported in May that children would be safe at schools located within a 60 km radius of Daiichi and that exposure to radiation would be minimal. However, in June, the government extended the evacuation zone to include some spots which were showing increased levels of radiation. Residents in these areas, although not obliged to leave, were urged to consider the risks, especially if they had children.

Some residents in the extended evacuation zones went for full-body radiation checks. Parents' concerns that their children's schools were contaminated by radiation and that it was unsafe for their children to return in September after the summer holidays prompted local authorities to issue radiation dosimeters to 34,000 children to measure the levels of radioactivity in their thyroids over a three-month period. The government submitted a report to the Japan Atomic Energy Agency (JAEA) on 7 June stating that 200,000 people in Fukushima had been screened and that no health problems were detected. However, low-level radioactivity was found in 1,000 children[8] and in May traces of cesium-134 and 137 were detected in the urine of ten children. The Fukushima government began a thirty-year project to test for internal exposure to radiation and detected cesium-137 in the urine of fifteen residents who lived in towns within a 25 km radius of Daiichi.

Consequences of the nuclear crisis: Fukushima Daiichi workers

The limit for radiation exposure for nuclear plant employees is 20 mSv a year or 100mSV over a period of five years. Even though workers were rotated to reduce time spent in the reactors, they could not escape significant exposure while struggling to bring the crippled plant under control. Twenty-eight of them received high doses of radiation. Three workers were contaminated by very high levels of radiation equaling 170–180 mSv when they laid cables in the basement of No. 3 reactor's turbine building. They were later hospitalized when they showed signs of radiation poisoning. One woman employee working at the plant after the earthquake tsunami was exposed to 17.55 mSv of radiation, three times the legal limit for workers.

Regardless of the high risks, on 15 March the MHLW announced that the limit considered safe for exposure to radiation would be raised from 100 mSv to 250 mSv after receiving a request from the Ministry of Economy, Trade and Industry (METI), which oversees the power utilities, to raise the limit to deal with the emergency. METI made the decision after consulting with the Ministry of Education, Culture, Sports, Science and Technology (MEXT).

TEPCO refused to accept the new limit and announced that, as of 1 April, workers had already been exposed to a cumulative dose exceeding 100 mSv. TEPCO had set the target rate at the plant at 100 mSv but had set lower limits of 80 mSv a year in order to control radiation exposure. TEPCO admitted that it had complied with the higher limit but had to reduce the levels because not all workers were carrying radiation monitors with alarms due to short supplies. Construction companies on the site also adopted the 100 mSv a year limit.

On 25 April, the MHLW issued instructions to TEPCO and to the subcontractors to send their employees for medical examinations.

On 7 May, TEPCO confirmed that, although thirty workers had been exposed to more than 100 mSv of cumulative radiation since 12 March, they had not received examinations until May. This was almost two months after the crisis began. Furthermore, 800 workers who had spent more than a month dealing with the nuclear crisis were just beginning to be examined by doctors commissioned by TEPCO and by companies that were subcontracted to assist TEPCO workers.

On 30 May, TEPCO reported that two of its workers had been exposed to well over the government limit of 250 mSv since 11 March but were not showing any evidence of illness. NISA reported that the workers had absorbed over 600 mSv of radiation. Following an on-site inspection, the MHLW reported that the two men in question had worked in the central control room of No. 3 and No.4 reactors and in a building where the crisis control center was located after the disaster hit the complex. The ministry ordered TEPCO to improve working conditions at the plant.

On 14 June, TEPCO confirmed that eight more workers were found to have been exposed to more than 250 mSv of radiation and that, altogether, 102 workers had received doses of radiation over 100 mSv since 11 March. TEPCO later disclosed that 1,600 workers had received doses of radiation exceeding 50mSv since 11 March. NISA announced that it intended to raise the 50 mSv limit not only for Daiichi workers but also for staff at all of the plants.

Although Japan has 54 commercial nuclear power plants that supply 30 percent of the country's entire power output, the Japanese have no expertise in decommissioning plants with problem reactors and decommissioning four reactors is far more difficult than for one reactor, which was the case for both Three Mile Island and Chernobyl. To further complicate matters, Fukushima Daiichi had three active reactors that were not cooling properly. The plant, built in 1971, is one of the oldest in Japan and, according to TEPCO, there were differences in the degree of safety levels between Daiichi and the newer plants regarding power source equipment such as emergency diesel generators, transformers at the reactor cores and pumps for pumping in seawater to remove residual heat from the cores. Toshiba and Hitachi had installed emergency generators and pumps in the newer plants but similar equipment was not installed in Daiichi.

Toshiba Corp., which supplied four of Fukushima's six reactors, including two with General Electric Co., consulted about plans to decommission the crippled reactor. The conclusion was that it would take approximately ten years to remove the fuel rods and the reactors and contain radioactivity at the site. Although TEPCO promised in April that the plant would be in cold-shutdown within six months, by August as radioactivity continued to leak from Daiichi, it was considered highly unlikely that the plant would be brought under control by the end of the year.

'Information-sharing' is not a buzz-word in government agencies

It will take months, even years, to assess the effects of the radioactivity on residents' health. Japanese government officials criticized the foreign press for exacerbating public concerns by misrepresenting facts related to the radioactive fallout. But since the foreign press were allowed access only to government agencies they were assigned to cover, it was a struggle to get reliable information. Wolgar Weiss, chairman of the UN Scientific Committee on the Effects of Atomic Radiation, told the press on 7 April that the committee was not getting the information necessary to enable an estimation of the consequences of the release of radioactivity and that Fukushima was considered a far more serious accident than Three Mile Island.

The foreign press was not the only victim of the ministries' reticence to release data. Prime Minister Kan was unaware of the seriousness of the damage to the reactors until the evening of 11 March. He was infuriated that he had not been informed as soon as TEPCO and NISA became aware of an impending crisis.

The earthquake struck at 2.46 p.m. NISA officials claimed that they were first alerted by TEPCO at 3.45 p.m. that AC power sources at Fukushima's No. 1, 2, 3, 4 and 5 reactors were lost and that only the B emergency generator at No. 6 was working. All electric power had shut down in the central control rooms and TEPCO officials recognized that the reactor cores could not be cooled without power sources. At 5.45 p.m. NISA official Koichi Nakamura told reporters that, although water was being pumped into the reactor cores, they did not know the level of the water. TEPCO officials issued a report to the Prime Minister's Official Residence stating that no problems were expected for eight hours because emergency batteries could be used for eight hours to substitute for AC sources.

Haruki Madarame, chairman of the Nuclear Safety Commission of Japan (NSC) informed Kan that there was no radiation leaking into the atmosphere and that the only existing problem was cooling the reactors.

At 5 p.m. Kan told the Japanese people that, although Daiichi had stopped, there was no confirmation that radioactive materials were escaping into the atmosphere. However, at 7.45 p.m. a state of emergency was declared. By 1.30 a.m. on 12 March, Madarame and TEPCO officials visited the Prime Minister's residence to report that the pressure was rising in No. 1 reactor and that a large volume of steam had accumulated in the reactor's containment vessel. Kan decided to see for himself what was transpiring at Daiichi and flew to the plant at 7 a.m. the next morning. He demanded a full explanation and urged that ventilation of the pressure be initiated immediately. Masao Yoshida, the plant's director, assured Kan that the situation would be handled appropriately. Subsequently, the prime minister dealt directly with Yoshida which effectively cut off communication with TEPCO's headquarters in Tokyo.

Kan was vulnerable to communication gaps with the ministries. He had been tackling Japan's ministerial administrative system since 1998. Kan first made headlines in the Japanese press while he was the Minister of Health and Welfare in Ryutaro Hashimoto's coalition government in 1996. When he investigated the ministry for its collusion in the distribution of blood tainted with HIV to hemophiliacs officials gave him the information he needed but on the condition that he would not release it to the public. However, Kan ignored their request. After he left office his information pipeline was cut and news regarding the investigation never went beyond the ministry's confines again.

Institutional reforms postponed

Kan, one of the founders of the Democratic Party of Japan (DPJ) in 1998, continued to pursue institutional reforms of the bureaucracy with fellow DPJ politicians. In 2001, the DPJ manifesto called for the elimination of the *amakudari*. Japan's post-war civil service system includes a system known as *amakudari* ('descent from heaven') which is an institutionalized practice through which the National Personnel Authority (NPA) formally places bureaucrats who have reached retirement age into private and public organizations. The arrangement provides retired officials with re-employment in top-level management positions where salaries are often higher than their previous incomes. Many positions are in the corporations whose industries are within the officials' ministries' administrative jurisdiction. The officials have a direct link to their former ministries and, therefore, can lobby in support of their employers' interests and concerns. The companies reciprocate by acquiescing to ministerial policies thus enhancing the ministries' regulatory control.

Some of the jobs are in the power utilities, which METI regulates. As an example of METI officials migrating to TEPCO, Toru Ishida, the former director-general of METI's Agency of Natural Resources and

Energy Agency (ANRE) was employed in January.[9] On 13 April, Chief Cabinet Secretary Yukio Edano announced that *amakudari* was 'socially unacceptable' and requested that senior METI officials in the NISA and ANRE temporarily abstain from taking positions in power companies 'in order not to raise the public's suspicions'. It was evident that the word 'temporarily' signified that post-retirement positions for METI officials in the utilities were likely to continue.

Amakudari is a primary component in Japan Inc. It was widely believed in the 1970s and 80s that the interlocking relationship between business, bureaucracy and politicians was the secret of Japan's success in global markets. The system seemed to work well while Japan was emerging industrially. However, what was not generally recognized was how the interlocking relationships created a mutuality of vested interests among the stakeholders, a very insular environment that protected domestic businesses from foreign competition and a rigid system of ministerial administration of the economy, a model that would ultimately defy efforts since the 1990s to implement structural reforms.

Although the practice has come to be regarded as encouraging corruption and sloppy regulation, it was not until the beginning of the 1990s after the asset-inflated bubble burst and scandals emerged involving *amakudari* and the ministries' collusion with companies in their administrative jurisdiction that *amakudari* was no longer publicly condoned. Despite some reform measures initiated in 1998 and 2001, the system continued.

In 2009, after the DPJ won the majority of seats in the Lower House, which had been dominated by the Liberal Democratic Party of Japan (LDP) since 1955, it called again for the eradication of *amakudari*. But the ministries, who are in charge of implementing the reform, procrastinated because not only did the system provide positions in the private sector but, as importantly, the system offered easy migration to jobs in their Independent Administrative Institutions (IAI) (known until 2003 as Special Corporations), industrial associations and research institutes where the number of jobs vastly outnumbered those in the private sector. Furthermore, jobs in these organizations enabled officials to migrate smoothly to positions offered by the private sector.

By trying to bust bureaucratic rule Kan and the DPJ became the odd-men-out and, consequently, their policies were not enthusiastically supported by the ministries. Indeed, Kan's administration finally came to terms with the fact that in order to implement economic reforms and trade agreements, it had to rely on ministerial support and began back-tracking on promises to weaken the bureaucracy. For example, the reform initiated by the DPJ in September 2009 that ended the system of agenda-setting meetings with the ministries' vice-ministers and limited contact between bureaucrats and politicians received a blow in December 2010 when Chief Cabinet Secretary Yoshihito Sengoku promoted the reinstatement of meetings between vice-ministers and deputy vice-ministers and politicians holding the top three ministerial posts in the cabinet in order to forge better communication regarding policy. In other words, Kan tried to re-establish his pipeline to the ministries.

But on 11 March, Kan, who was well known for his deep mistrust of bureaucrats, considered the nuclear energy industrial sector to be embedded in a web of collusion between METI bureaucrats, the utility companies, politicians and submissive academics. His reliance on only close colleagues to advise him on the escalating crisis isolated him from the bureaucrats and politicians who were privy to more information about the conditions at Daiichi. Kan's indecisiveness in dealing with Daiichi and the prolonged effect of the crisis on Japan's economy culminated in Kan's resignation the following August.

Fukushima Daiichi Nuclear Power Plant: an impending disaster

Scientists are still in the process of creating technologies that will correctly forecast the exact time an earthquake will strike a specific region. However, when nuclear power plants are constructed in close proximately to seismic fault lines there is good reason to believe that when an earthquake does strike, the power plant will sustain considerable damage. Katsuhiko Ishibashi is a professor in the Research Center for Urban Safety and Security at Kobe University and one of Japan's preeminent seismologists. He opposes the construction of nuclear power plants in Japan and his views are supported by a number of Japanese seismologists. Speaking to the Committee on Audit Administration in the National Diet on 23 May, Professor Ishibashi stated that there was a high probability that the nuclear accident at Daiichi had been caused by the earthquake tremors before the tsunami hit. He considered that the earthquake had damaged the pipeline in No. 1 reactor carrying the coolant and therefore the cooling ability was lost. Ishibashi told the committee that the strong shocks could have damaged the suppression pool in No. 2 reactor, causing it to lose its function. Furthermore, the foundation boards in No. 2, No. 3 and No. 5 reactors were not built to withstand either the magnitude of the quake or the duration of the shocks.

Daiichi had experienced problems since it was commissioned in 1971. After the Great Hanshin Earthquake in 1995, Jinzaburo Takagi, the late former director of the Citizens' Nuclear Information Center and an associate professor at Tokyo Metropolitan University, published a paper entitled 'Nuclear Facilities and Emergencies, with Focus on Measures against Earthquakes' in the October 1995 issue of the *Physical Society of Japan*. The paper attacked the government and the power companies for not including emergency measures at nuclear power plants in the event of earthquakes and tsunamis because they had presumed that the power plants were fail-safe in earthquakes. Takagi pronounced the Great Hanshin Earthquake in January 1995 as a 'wake-up call' for preparing nuclear power stations for emergencies such as earthquakes and tsunamis. He referred to Fukushima Prefecture's Hamadori coast where there were a number of nuclear facilities claiming that, if an earthquake struck, the region would encounter an emergency that would be beyond imagination. Takagi expressed great concern about Daiichi because it was very old and should have been decommissioned.[10]

An investigation of the disaster will pose questions regarding why Daiichi was built in an earthquake-prone zone, why it was still operating despite problems such as cracks in infrastructure and leaks of radioactivity, an incident that killed one worker, complaints filed by former Fukushima governor Eisaku Sato in 1989 and 2002, and why the same types of safety equipment that existed in newer plants had not been installed in Daiichi.

Eisaku Sato served five terms as governor of Fukushima from 1988 to 2006. He was forced to resign when he was convicted by the Tokyo District Court for his participation in a bribery case involving a public works project. He received a suspended prison sentence in 2008 and the verdict was upheld in 2009. He is currently appealing to the Supreme Court.

Sato told the press on 7 April that in January 1989 he became aware that TEPCO's maintenance of Fukushima's nuclear power plants was incompetent. TEPCO, knowing that one of the reactors' cooling pumps at Fukushima's second power plant was not operating up to standard, failed to report it to the prefectural authorities. The second power plant, which was built after Daiichi, was fitted with more modern safety equipment. Sato quickly filed a formal complaint with the Ministry of International Trade and Industry (MITI) (renamed the Ministry of Economy, Trade and Industry, METI in 2001). However, ministry officials did nothing more than to admonish TEPCO and the plant was operating again after a temporary shutdown.

In 2002, Sato filed a second complaint with METI when a whistleblower maintained that TEPCO was concealing malfunctions and cracks at both Daiichi and Fukushima's No. 2 power plant. The agency that regulated the nuclear sector prior to NISA's establishment evidently had received the same information in 2000 but the former MITI chose not only to ignore it but also did not inform the prefectural government authorities. According to a NISA official, the agency passed it on to TEPCO. Sato's complaint triggered the resignation of some of TEPCO's management and a temporary shutdown of all of TEPCO's nuclear reactors.

During the remainder of Sato's incumbency the prefectural government received twenty-one anonymous tips of a broken turbine that TEPCO did not report as well as the shortage of safety measures in the facilities. Sato blamed the crisis at Daiichi on the close relationship of METI officials with the utilities, resulting in the cover-up of major problems in the reactors.

METI issued instructions in 2006 to electric power utilities to perform comprehensive inspections, sending an individual instruction to TEPCO. However, Niigata Prefecture governor Hirohiko Izumida was extremely concerned about the disclosure in 2002 that TEPCO had submitted to government falsified safety inspection reports regarding Daiichi. On 15 February 2007, he personally visited the office of Akira Amari, Cabinet Minister of Economy, Trade and Industry, to deliver a letter requesting that NISA ensure the safety and security at power plants, specifically Kashiwazaki-Kariwa, in Niigata. This, the largest nuclear power plant in the world, had sustained considerable damage in the Niigata Chuetsu-Oki Earthquake a month earlier and had been temporarily shut down. Minister Amari assured the governor that METI would wipe out past corruption, investigate the cause of the problems at TEPCO and ensure complete disclosure.

At his news conference in April Sato urged the separation of NISA from METI. The prime minister also was in favor of the separation.

Cover-ups of faulty equipment and flaws in the design of infrastructure at Daiichi can be traced back to 1974. Mitsuhiko Tanaka was an engineer at Hitachi-Babcock KK and helped to design and to supervise the manufacturing of a $250 million steel pressure vessel for TEPCO in 1975, which was damaged during production. The vessel holds the fuel rods in Daiichi's No. 4 reactor. Tanaka published a book in 1990 called *Why Nuclear Power is Dangerous*, in which he acknowledged his part in arranging a cover-up regarding the damaged vessel, which saved his company millions of dollars. He was rewarded with a $38,000 bonus. Two years after Chernobyl, he went to METI to report the cover-up. Hitachi denied his accusations and METI refused to investigate further.[11]

Where does the buck stop?

Before the end of the Second World War the Japan Electric Power Company provided electricity to the entire nation. The monopoly was broken up into smaller electric companies by the Supreme Commander of the Allied Powers (SCAP). TEPCO, established in 1951, is the largest among the ten electric utility companies, the largest in Asia and the fourth largest in the world. It controls one-third of Japan's electricity market. It purchases electricity from wholesalers, including Tohoku Electric Power Co., J-POWER, and Japan Atomic Power. TEPCO received much of the blame for Daiichi but its close ties with METI implicate the ministry as well because METI administrates both the nuclear power industry and the power utilities.

J-POWER was known until 2004 as Electric Power Development Company (EPDC), a public corporation established in 1952 under the former MITI's administration. The agency managed all of the electric power utility companies. EPDC was privatized in 2004 and renamed J-POWER, as a METI affiliated organization. In addition to coal and hydroelectric power stations, J-POWER is building a nuclear power plant in the town of Oma in Aomori Prefecture. During Japan's post-war period, MITI officials received post-retirement positions in EPDC and in its subsidiaries, which enabled them to establish close contact with the electric utility companies and to migrate to upper management positions in the companies.

Since the late 1990s *amakudari* in the private sector has been on the decrease as corporations have become disillusioned with government failure to resolve Japan's numerous economic problems. Some younger officials, disenchanted with Japan's continuous economic malaise, the hierarchical and insular structure of the ministries and the deterioration of the prospects of receiving well-paid jobs at the end of their careers, were leaving the ministries mid-career to seek employment in the private sector.

As jobs in the private sector are decreasing, the ministries' Independent Administrative Institutions (IAI), research institutes and industrial associations are becoming more significant as full-time employers. It was reported that during the period from September 2009 to 4 October 2010, 240 retiring ministry officials migrated to jobs in IAIs, other ministry-related organizations such as research institutes and industrial associations and the utilities; 679 officials assumed positions on the boards of these organizations.

IAIs were known as Special Corporations before 2003. Established by the national ministries after the Second World War to aid in the reconstruction of infrastructure destroyed during the war and to resuscitate industry, the corporations are linked to the industrial sector under the administrative jurisdiction of each ministry and are supported by public finance. After the war a large portion of the fiscal stimulus packages went towards subsidizing the construction of infrastructure such as roads, bridges, dams and housing and a number of the corporations distributed contracts to private corporations.

IAIs can function as a means for officials in post-retirement positions to establish close relationships with private corporations thus promoting employment opportunities in the businesses. Since the ministries in charge of public works projects will engage with businesses through their IAIs, businesses who are eager to win contracts may offer employment to officials posted in the IAIs in exchange for information regarding the lowest price set for the contracts.

Special Corporations continued to operate out of the public eye until 2000 when revelations of corporations burdened with heavy debt due to mismanagement, together with court cases involving bid-rigging involving Special Corporations, prompted a strong political movement led by former Prime Minister Junichiro Koizumi to privatize or merge or dissolve the corporations. Two large debt-ridden major corporations were privatized while others underwent mergers. Koizumi renamed Special Corporations 'Independent Administrative Institutions' to signify that the organizations' operations would become more transparent and that there would be further privatization and mergers to reduce the number of ministry-affiliated institutions. Regardless of these efforts, the ministries were able to maintain many of their corporations because as Carpenter (2003) stated: 'To date, Koizumi's plans have met with stiff opposition because of the vested interests of the bureaucracy, businesses and the LDP.'[12]

METI established Special Corporations that had been engaged in the development of nuclear energy and nuclear reactors during Japan's post-war period. As an example, METI established the Japan Nuclear Cycle Development Institute (JNC) in 1998. Its projects included fast breeder reactors, a plutonium fuel fabrication plant and a MOX fabrication plant. In 2005, the JNC merged with the Japan Atomic Energy Research Institute (JAERI), an IAI of the Ministry of Education, Culture, Sports and Technology. Both institutes were involved in similar projects.

The *amakudari* system will undoubtedly be held responsible for the cover-ups and poor supervision of Fukushima Daiichi. However, the ministries' IAIs and research institutes and industrial associations are the vehicles that enable *amakudari* and they are still at the very heart of Japanese government. As extensions of the ministries, they are the mechanisms that connect the ministries to the private sector and, also, to local governments.

In 2002, the Organisation for Economic Co-operation and Development (OECD) called on the Japanese government to implement major structural reforms to reduce debt and get the economy out of deflation but, as Carpenter (2003) stated:

> Most reforms have been stymied because of fears that public funds that have continued to flow freely despite the recession will become a trickle. Political bickering among the LDP and among members of opposition parties [DPJ] has resulted in political gridlock. In fact, the philosophies of opposition parties which were once at opposite ends of the philosophical spectrum are now converging as the LDP platform encompasses their platforms. The Japanese are in political limbo.[13]

Former Prime Minister Junichiro Koizumi (2001–6) was a staunch member of the Liberal Democratic Party (LDP). He was determined to cut public spending and reform some of the IAIs that served to extend ministerial influence over private industry and, consequently, over the Executive as well. Ultimately, the reforms proved to be superficial at best and, indeed, the LDP lost seats in the National Diet because local businesses were concerned about the decrease in public works projects. The public debt continued spiraling to 140 percent of GDP.

After Koizumi resigned from office in 2006 he was followed in quick succession by Prime Ministers Shinzo Abe (September 2006–September 2007), Yasuo Fukuda (September 2007–September 2008) and Taro Aso (September 2008–September 2009) who were LDP politicians. They struggled to stabilize the economy, end deflation and rein in the public debt. They attempted to tackle the same reform issues as Koizumi during their short terms in office. On the other hand, they adhered to the same fiscal and monetary policies. Japan's rigid and insular model of government administration continued to frustrate their efforts to pass reform bills in the Diet. Carpenter (2008) stated that:

> Pressure from the IMF and OECD to continue reforms will not propel government to pick up the pace because it cannot. The Japanese are mired in a system of administration that is rooted in interpersonal relationships between an elite bureaucracy, a dominant political party and big business.[14]

The same old story

The DPJ's landslide victory in the Lower House election on 30 August 2009 succeeded in breaking up the power of the LDP. The electorate had high hopes that the DPJ would bring about a transformation of government administration through reforms that it had promoted in its manifesto, including cutting the public debt through fiscal policies, initiating institutional reforms of the bureaucracy and introducing child welfare allowance.

On 9 September, Yukio Hatoyama became Japan's first DPJ prime minister (September 2009–April 2010). He was compared to John F. Kennedy because of his youth and a wealthy family who supported his campaign with large contributions. A degree from Tokyo University and a PhD from Stanford University also added to his elite image. Unfortunately, he was no more successful than his LDP predecessors and resigned from office in April 2010 after only seven months in office amid political funding scandals involving DPJ kingpin Ichiro Ozawa. He also had been widely criticized about his indecisiveness regarding the relocation of the US Marine Corps Air Station on Okinawa, which has long been a highly contentious issue among the Japanese.[15]

Naoto Kan, Hatoyama's Cabinet Minister of Finance, took over in April and was elected president of the DPJ in June. When he assumed the office of Prime Minister on 8 June he promised to tackle Japan's beleaguered economy through fiscal reforms and cutting the sovereign debt which was over 200 percent of GDP, or twice the annual tax revenue. His cabinet's approval rating was 64.8 percent.

In May, the IMF asked Japan to cut its massive debt and recommended tax hikes as a way forward because, although Japan's economy was growing by 2 percent, it would stall if the debt, which was eating away at growth, was not brought under control.

It was predicted that the DPJ would expand its small majority in the Upper House in the July elections but when Kan indicated that his fiscal policies included raising taxes, his public approval rating plummeted to 36.3 percent. The LDP managed to scoop up some rural votes and block the DPJ's dominance in the Upper House elections.

Kan's cabinet's approval rating continued to fall. Within three months his administration had lost its glow due to scandals involving Ozawa and other key party officials who refused to admit to any wrongdoing, and Kan's inability to bring consensus. In January 2011, he reshuffled the cabinet to raise ratings in order to cling to his diminishing power-base. However, it appeared that Kan was in an untenable situation because along with the mounting public debt and the appreciation of the yen, the problems that he was trying to address had become too complex. Directly prior to the earthquake, Kan's approval rating had slipped to 17 percent and it was assumed that he would be pressured to resign from office. But the triple disaster on 11 March gave his administration a reprieve.

At the end of March Kan tried to persuade opposition parties, including the LDP, to form a coalition government to collaborate in the enormous task of the reconstruction of Japan's northeast. However, the LDP insisted that Kan step down as prime minister, indicating that the party was anticipating a come-back. Despite pressure from the majority of DPJ officials to tender his resignation in June in order to placate the opposition parties and pass key bills to support the enormous recovery costs, Kan refused to exit gracefully.

Kan's public approval rating slid to 15 percent in July but Kan managed to hang on to power until the end of August, using the crisis to try to promote policies he had wanted to implement prior to the earthquake. He refused to step down until bills calling for a ¥2 trillion secondary budget, a ¥10 trillion third budget supported by deficit-covering bonds for the post-crisis recovery, and a shift from nuclear to renewable energy were passed by the Diet.

In early July, Kan demanded new safety regulations and inspection tests of all of the idled reactors without consulting members of his cabinet. His unilateral action not only created a major disruption of power supplies in Japan during the hot summer months but further alienated opposition parties and businesses, bringing the DPJ, the party he had founded, to its knees. The LDP countered with a pledge to keep the power plants operating while improving safety standards and expanding public works projects for disaster control. The government was in chaos and the Japanese were falling into a political abyss.

Four months after the 11 March disaster Kan's government was bowing to pressure from the LDP to backtrack on its 2009 manifesto pledges in order to pass the third supplementary budget. On 22 July, despite protests from DPJ members, the party formally acquiesced to LDP demands and withdrew its child welfare allowance and other key pledges put forward in its manifesto. In anticipation that a new administration would be formed as soon as Kan vacated office the Diet passed two bills in August and the prime minister finally resigned. His cabinet minister of finance Yoshihiko Noda was elected prime minister on 30 August.

The routes to responsibility

On 13 March two days after the government announced the conditions at Daiichi as an 'emergency', TEPCO's president Masataka Shimizu disappeared from public scrutiny and entered a Tokyo hospital suffering from stress and high blood pressure. A month later on 13 April when the government announced that Fukushima had reached a 'level 7' Shimizu reappeared and visited the plant for the first time. The governor of Fukushima refused to meet with him before his press conference where he made a public apology. Shimizu assured the public that TEPCO was convinced that it had built nuclear power plants that could withstand natural disasters. He also confirmed that he was in perfect health and that it was back to work as usual. Shimizu resigned as president on 20 May when TEPCO reported an initial loss of ¥1.2 trillion ($15 billion) which excluded the decommissioning of the four reactors. TEPCO Managing Director Toshio Nishizawa stepped in as president.

TEPCO was not the only electric power utility company to release false reports. Kansai Electric Company (KEPCO) also admitted to falsifying reports. Following Shimizu's resignation as the chairman of the Federation of Electric Power Companies (FEPC) in March, KEPCO president Makoto Yagi succeeded him. Yagi stated at a press conference in April that nuclear power remained a valuable and indispensable power source. In June, Shosuke Mori, KEPCO's chairman who also was the head of the Kansai Economic Federation, the biggest business lobby in western Japan, told reporters that nuclear energy was Japan's main energy source and that it was the only way to ensure a stable supply of environmentally clean electricity.

Perceptions of Japan Inc.: busting myths

The revelations from the ongoing investigation of Daiichi will inevitably effect a change in the way the international community has perceived Japanese government administration of the political economy. When the United States was in recession in 1981 Japanese electronics and automobiles were flooding global markets. The American Dream model of free-market capitalism, once considered the ideal model for economic development, was no longer in vogue. The international business community saw a new model, the Japan Inc. model which appeared to be the driver of Japan's emergence as a major technological and industrial force. Prior to 11 March, despite fiscal policies that focused on the continuous release of stimulus packages and supplementary budgets, resulting in a burgeoning public debt, and despite the Bank of Japan's soft monetary policies as well as the efforts by successive administrations to implement structural reforms of the economy, Japan was stuck in deflation. Nevertheless, it was still widely believed that the Japanese government would be able to resolve the problems that had been inhibiting stable economic growth and dig Japan out of fifteen years of deflation.

Like his predecessors, Kan was controlled by a rigid and inward model of administration, which was dysfunctional in dealing with the mounting problems Japan had faced since 1990. The efforts of law-makers to reform the institutions and institutional arrangements that have served to keep the model in place have progressed at a snail's pace. And there is no substitute for the current model that has paralyzed the policy-making process and prevented the Japanese from accommodating the ever-changing demands on the domestic economy and dealing with the instability of foreign markets. The Japanese must depend on this system to resuscitate an economy that has experienced the worst crisis since the end of the Second World War. They fully recognize that the model is archaic and inward but, ironically, they also say that it could take something as significant as a major disaster to compel reforms.

The challenges that the Japanese were facing prior to the triple disaster were numerous. External pressures on the economy were also increasing. Even if some reforms are implemented it may be too late to bring Japan's economy back on track. Japanese political society is conservative and risk-averse and reforms may be put aside in favor of the familiar status quo.

The route to responsibility for Fukushima leads to TEPCO and the power utility industry but it certainly does not stop there. It goes all the way to the heart of Japanese government which is the institutions and the institutional arrangements that serve to forge strong interpersonal relationships between government agencies and the private sector and nurture an environment of vested interests and mutual obligation. The route to responsibility ends at the center of Japan Inc.

This book provides a realistic account of Japan's political economic landscape and the elements embedded in it that have engendered a system which supports and sustains the ministries' organizations and institutional arrangements but simultaneously impedes the reform of the system. Besides NISA and the ANRE, METI IAIs, affiliated organizations and research institutes were also directly involved in the post-war development of nuclear energy and in the nuclear power and electric power industrial sectors and, therefore, implicitly involved in the nuclear crisis. The examination of these and other ministerial IAIs will reveal how the corporations can perpetuate the interests of the ministries; how, as extensions of the ministries, they can influence industry and local government; how the ministries can revise operations in order to maintain them; and how the ministry officials can access employment through the manipulation of *amakudari*.

1
Independent Administrative Institutions: In Name Only

The national ministries' Independent Administrative Institutions (IAIs) were known until 2003 as Special Corporations (*tokushuhoujin*). The corporations were established after the Second World War to aid in the reconstruction of infrastructure destroyed during the war to resuscitate Japan's industry. The ministries began to establish public corporations in 1947 during the Occupation at the encouragement of the Supreme Commander of the Allied Powers (SCAP). Four of the corporations were designated for foreign trade, eight supported domestic distribution, one served to control price adjustment and two were for economic rehabilitation.[1]

The government had difficulty defining the exact characteristics of Special Corporations other than that they assisted the government in promoting national interests. Nevertheless, they can be viewed as corporations that were based on a national law, which had been approved by the National Diet. Special Corporations were established according to special establishing procedures through a special law, the Law of Establishment Act, Article 4-11 subject to the Ministry of General Affairs (renamed in 2001 the Ministry of Public Management, Home Affairs, Post and Telecommunications). The law was neither civil nor corporate.

Special Corporations received funds through loans and investment from the Fiscal Investment and Loan Program (FILP). FILP was established in 1953 as a huge financial organ operated by the public sector to provide loans through FILP Agency Bonds to Special Corporations for contracts for public works, loans for infrastructure, housing, small and medium-size companies, mortgages and life insurance. FILP was financed by deposits in the Postal Savings System (PSS) and by premiums from the public pension scheme.

The corporations were linked to the industrial sectors under the administrative jurisdiction of each ministry. For example, the former Ministry of Construction (MOC) (renamed in 2001 the Ministry of Land, Infrastructure, Transport and Tourism (MLIT)) established the Japan Highway Corporation (JH) in 1956 to award contracts to construction companies to rebuild highway networks.

The former Ministry of International Trade and Industry (MITI) (renamed the Ministry of Economy, Trade and Industry in 2001) established the Japan Electric Power Development Co. (EPDC) in 1952 to provide electricity nationally. It established the New Energy Development Organization (NEDO) in 1980 for the research and development of industrial, energy and environmental technologies. MITI established the Japan External Trade Organization (JETRO) in 1956 to promote small business exports to overseas markets. The Ministry of Finance (MOF) established the Japan Development Bank (JDB) in 1953 to aid Japan's economic recovery.

By the early 1990s the ministries had established over one hundred Special Corporations. It was a simple procedure because all that was needed to establish a new one was the writing of an 'establish law' in the name of the corporation and a request to an obliging Diet to sanction it. Throughout the post-war era numerous subsidiaries ('children corporations') and subsidiaries of these subsidiaries ('grandchildren corporations') were established by ministry officials. Special Corporations, their 'children' and 'grandchildren' corporations and branch offices served to place officials from the national ministries throughout Japan. The EPDC's subsidiaries included the EPDC Environmental Engineering Services Co.; EPDC Coal Tech and Marine Co.; Kaihatsu Co.; Kaihatsu Computing Service; Kaihatsu Denki Co.; KEC Corporation; KDC Engineering Co.; and EPDC Overseas Coal Co. The EPDC also had a holding company and an industrial company. The branches served to provide posts for local government officials as well. Some of the corporations also operated representative offices overseas.

The National Public Service Law stipulates that bureaucrats cannot, for a period of two years, legally move directly to positions in private companies attached to the sectors their ministries regulate. However, they can move immediately to Special Corporations, industrial associations or research institutes supported by their ministries or by the other ministries where they linger for two years with pay before going on to the private sector. Most officials will wait out the two-year period of grace in public corporations and then slip into higher management positions in private companies (i.e. *amakudari*).

Ironically, when SCAP permitted the formation of a public corporation to support the development of the match industry, representatives from the industry, fearing tight controls over their sector, issued a petition on 10 August 1947 to SCAP in protest. The petition claimed that the corporation was being formed not to support the industry but to provide jobs for bureaucrats. SCAP rejected the petition because it assumed that the industry did not want ministerial controls. Later reflecting on the decision, SCAP believed that public corporations could resuscitate Japan's wartime system of autocratic economic controls.[2]

Until the asset-inflated economic bubble burst in 1990 it was believed that elite civil servants had created an economy that was impervious to failure. The Japanese generally accepted *amakudari* because it was one of the incentives that attracted talent into the ministries. However, from 1992 onwards the stagnation of Japan's economic growth, coupled with the revelation of collusion between the Ministry of Construction (MOC) and the construction industry, and between the Ministry of Finance and the banking industry triggered a growing demand from opposition parties for the dismantling of Special Corporations and the eradication of *amakudari*.

Since Special Corporations were seen as vehicles for the migration of officials to upper management positions in private corporations, the media began focusing attention on the ministries' corporations. In February 1997, Japan's largest business weekly *Nikkei Business* interviewed Hiroshi Kato for its cover story on Special Corporations. Kato, who was the President of Chiba Commercial College and the chairman of the government's Tax Commission in the 1980s, emphasized that public funding of Special Corporations was a serious problem because the ministries had the power to use the money at their discretion without seeking consent from the Diet. Indeed, the politicians supported this behaviour because they solicited contracts from the corporations involved in public works for their constituencies.

Kato complained that there was no public disclosure by Special Corporations for accounts indicating profit–loss balances and that the accounting system used was difficult to fathom because it differed from the system used by private corporations. He recommended privatization. Some of the corporations had already been dismantled prior to 1996, when the number stood at ninety-two.

Nikkei Business claimed that while Special Corporations had been founded on the precept that the work executed would serve the national interest the opposite was true for the following reasons.

(i) Special Corporations received funding from sources that were difficult to trace.
(ii) Special Corporations could set up subsidiaries ('children' and 'grandchildren' corporations) that showed profits even though the parent corporations were in debt.
(iii) The ministries established Special Corporations as their subsidiaries to provide temporary employment for staff and post-retirement positions for retired senior officials before they moved to the private sector.
(iv) Special Corporations used funds to do work that was in the best interests of the corporations.

Prior to 1999 the ministries were not obliged to make public the budgets or operational expenses of their organizations. However, in 1999, the Law Concerning Access to Information Held by Administrative Organs was enacted. Article 42, Information Disclosure by Independent Administrative Institutions and Public Corporations, stipulates:

> In accordance with their character and type of business, the government shall take necessary measures such as legislative measures relating to the disclosure and provision of information held by independent administrative institutions and public corporations.

As of October 1999 the number of corporations had been reduced to eighty-four. Some Special Corporations began opening their books to reveal more than had been available before 1999. However, many of the corporations, including their subsidiaries, did not provide financial statements.

The push for reforms: the perfect opportunity

In 1992, the recession and the acknowledged need for action to ignite Japan's lagging economy triggered a struggle for power among factions in the Liberal Democratic Party, resulting in some members defecting to form new parties. Morihiro Hosokawa, who formed the small reformist New Japan Party, joined forces with the Japan Renewal Party, which was founded by Ichiro Ozawa. In the summer of 1993, the LDP lost its majority in the National Diet for the first time in thirty-eight years. An eight-party coalition government assumed power and Hosokawa became prime minister. There were high expectations that the new government would tackle the recession and initiate political reforms. The government tried to jump-start the economy by releasing several fiscal stimulus packages. A major portion of the funds was targeted for public works projects to provide contracts to businesses in the prefectures and to keep unemployment figures low. However, after serving only eight months in office Hosokawa was forced to resign amidst rumors that he had accepted a loan from a parcel delivery service whose previous loans to other politicians had caused political scandal.

A political struggle for power between the new parties ensued and Tsutomu Hata, who was the vice-prime minister in the Hosokawa cabinet stepped in as prime minister in April 1994 until Tomiichi Murayama, a Socialist, assumed the post in July, forming another coalition cabinet composed of members of the Socialist Party of Japan, New Japan Party, and the LDP. He began the push for a review of bureaucratic powers and attempted to reform the political system itself in order to end the relationship between politicians, bureaucrats and businessmen. This movement prompted a power struggle between politicians and bureaucrats, who were intent on maintaining control over the regulation of the economy.

In 1995, Murayama's efforts to convince the ministries to consolidate some of the smaller Special Corporations attracted media attention. The Opinion Page of the Internet edition of *Asahi Shimbun* ran an article on 9 January reporting that the former Ministry of Agriculture and Forestry and Fisheries balked at dissolving its Raw Silk and Sugar Price Stabilization Corporation and the Livestock Industry Promotion Corporation. The ministry contended that the corporations would continue to protect consumers by planning strategies that would stabilize prices. However, it was willing to merge the two corporations and in 1996 the entities were united and named the Agriculture and Livestock Industries Corporation (ALIC).[3]

Murayama's administration suffered from incidents that were remarkably similar to what was besetting the DPJ. There was the Kobe Earthquake on 17 January 1995, which brought harsh criticism of the government's handling of the aftermath, the continuing recession, political infighting and the opposition of Socialist Party members to Murayama's support of the Japan–US Security Pact which Murayama contended was in accordance with Japan's Constitution.[4] Furthermore, the Sarin gas attack by the religious cult Aum Shinrikyo in the Tokyo Underground on 20 March 1995, which killed twelve people and injured 6,000, increased Japanese anxieties about the country's socio-economic stability. Voters returned to the LDP fold and the Socialist Party lost many seats in the Lower House in the 1996 elections. Murayama resigned as prime minister making way for the LDP president Ryutaro Hashimoto to enter the executive office. Frustrated members of the new parties either returned to the LDP or joined other opposition parties.

Nevertheless, Murayama's reform efforts opened the door to close scrutiny by the news media of the ministries' use of public corporations. The number of books and articles critical of public corporations increased as the recession continued. The editorial staff of *Mainichi Shimbun*, one of Japan's six major dailies, entered the fray in 1994 with a book focusing on *amakudari* in both private and public corporations. *The Kasumigaseki Syndrome*[5] (*Kasumigaseki Shindoromu*)[6] is a surprisingly frank account of the deterioration of values among bureaucrats in terms of their objectives in establishing corporations and research institutes for the sole purpose of providing post-retirement positions for elite retirees.

The *Mainichi* staff detailed how ministries maintained control over their sectors by placing conservative elite retirees into management positions in both private and public corporations. In addition, they revealed how the ministries used Special Corporations to distribute funds to companies to cover contracts for public works, and how *amakudari* not only tied ministries to businesses but also facilitated connections between businesses and former bureaucrats, who migrated first to Special Corporations before moving on to the private sector.

The book's most significant contribution was the reporting on how *amakudari* and the temporary posting of elite officials in the branch offices of Special Corporations in the prefectures (*shukko*) helped the ministries to monitor local government policies. Since the bureaucratic hierarchy places officers from the national ministries above local government officers, the positioning of ministry officials at the local government level automatically induces acquiescence by local government to ministerial guidance. The book was also critical of temporary postings because there was the ever-present possibility that the positions would become permanent.

During Prime Minister Hashimoto's coalition government administration (1996–2001) there was an effort to streamline government administration through the mergers of ministries and the integration of some of the minor agencies into the ministries and making some of the agencies that previously had been connected to the ministries into independent agencies. However, the mergers, rather than making government smaller, served no more than to change the names of various ministries and to create turmoil within the newly merged ministries regarding the restructuring of management of the institutions and territorial issues and objectives. In addition, officials were very concerned about how the mergers would impact on ranking and promotion.

In January 2001, the Ministry of Construction merged with the Ministry of Transport, the National Land Agency and the Hokkaido Development Agency to form the Ministry of Land, Infrastructure, Transport and Tourism. Prior to the merger, the Ministry of Transport had managed 849 public corporations. The Ministry of Education, which managed 1,811 public corporations, merged with the Science and Technology Agency to form the Ministry of Education, Culture, Sports, Science and Technology. The Ministry of Home Affairs merged with the Ministry of Management and Coordination and the Ministry of Posts and Telecommunications to form the Ministry of Public Management, Home Affairs, Posts and Telecommunications. The Ministry of Health and Welfare merged with the Ministry of Labor to form the Ministry of Health, Welfare and Labor. The Ministry of Finance retained autonomy as did the Ministry of Agriculture, Forestry and Fisheries, the Ministry of Justice and the Ministry of Foreign Affairs. MITI was renamed METI or the Ministry of Economy, Trade and Industry. The former Agency of the Environment was upgraded to the Ministry of Environment (MOE) and the Defense Agency was upgraded to the Ministry of Defense (MOD) on 9 January 2007.

The newly merged ministries managed jointly the Special Corporations which had been established and managed by a single ministry prior to the merger.

By April 2002 seven more corporations had been privatized or dismantled. Special Corporations that were big spenders but were burdened with debt were those involved in finance, such as the Long-Term Credit Bank of Japan (LTCB) which was established by MOF in the 1950s to finance industrial development.

After the collapse of the asset-price bubble the LTCB was riddled with \$19.2 billion of debt. After the Asian financial crisis in 1997, the debt burgeoned to \$50 billion. The government nationalized it and pumped \$2 billion into the failing bank. A consortium led by Ripplewood, a \$300 billion American private equity firm, purchased the bank in March 2000 for \$1 billion after the government was unable to find a suitable domestic bank for a merger. The contract called for government to pay at least \$45 billion to cover bank losses. The bank was renamed Shinsei ('new life'). LDP members in the Diet were not pleased with Ripplewood's purchase of the state bank. The media also expressed displeasure at the entry of foreign venture capital firms into Japan. When Ripplewood sold its shares in Shinsei in 2004 for \$2.4 billion and then sold more shares for \$3 billion in March 2005, earning over \$6 billion for its investors, there was public outcry since government investment in the bank to seal the deal was twice that of the American consortium. It was the first and last time that a foreign company purchased a state bank.

The ministries agreed to part with some of their corporations through the merger of insolvent with more solvent corporations. Two Special Corporations that MITI established in the 1950s – the Japan Export–Import Bank (EXIM) and the Bank for Overseas Economic Cooperation – were consolidated in 1998. The Hokkaido-Tohoku Development Finance Corporation, a bankrupt public corporation, established and managed by MOF, was merged with MOF's JDB in order to mop up its outstanding non-performing loans. The merger was celebrated with a new name, the Development Bank of Japan (DBJ). The DBJ was slated to privatize by 1 October 2008; however, the 2008 global financial crisis prompted the government to delay privatization. Since then, the DPJ has been pumping billions of dollars into bankrupt companies that were considered too big to fail, such as Japan Airlines. In June 2011, the bank joined together with the Japan Auto Parts Industries Association which represented 400 auto parts manufacturers and their main banks to create a fund initially worth ¥50 billion ($620 million) for the injection of capital into small subcomponent manufacturers involved in the auto supply chain in order to support the repairs of damage to existing installations and equipment caused by the 11 March earthquake.

Reforms according to Koizumi: FILP

Prime Minister Koizumi, an ultra-conservative member of the LDP, entered office in April 2001 with the pledge to slash public spending in order to cut government deficit, which was 130 percent of GDP at that time. Besides cutting public spending, Koizumi's other priorities were dealing with the non-performing loans problem and reforming the banking industry. He also intended to privatize the PSS because of the ambiguous methods used for dispersing funds to FILP. Koizumi was determined to downsize FILP and streamline Special Corporations through privatization or mergers. Many of the loans to FILP were not repaid and Seiji Ota, who was the director of the LDP office for the promotion of reforms, contended that one-third of the funds allotted to Special Corporations by FILP were wasted. Since MOF had been reticent to use the formal budget and tax revenue to finance the increasing number of loans throughout the 1990s, FILP was burdened with toxic debt.

During Prime Minister Ryutaro Hashimoto's administration efforts were made to implement structural reforms of FILP in order to stop the automatic supply of funds from the PSS and state pensions and halt indiscriminate and wasteful spending. FILP's ties to PSS and the pensions were cut in 2001 and neither body was required to fund FILP. Until 2001 FILP received its capital from the Trust Fund Bureau (TFB) where deposits from the PSS had been transferred for public financing. However, the TFB was terminated in 2001 and the PSS, as a part of its reorganization, was no longer obliged to transfer money to fund FILP. The Fiscal Loan Fund (FLF) replaced the TFB and Special Corporations were pressured to access capital by issuing their own bonds or FILP Bonds which were issued by the FLF. FLF Bonds were guaranteed by the government.[7] Corporations unable to access capital independently were funded through loans from FILP Agency Bonds. Even though the PSS was no longer required to invest in public corporations, since 2002 postal savings were used to purchase FILP Bonds and FILP Agency Bonds at the discretion of management.

Koizumi intended to downsize FILP by 17.7 percent or ¥26.79 trillion ($218 billion) but instead of diminishing in size, FILP expanded. During 2007–8 loans from FILP increased 14 percent and from 2008–9 loans increased 72 percent, clearly indicating that FILP remained 'Japan's secondary budget'.

Reforms according to Koizumi: enthusiastic support from the opposition

Koizumi's reform of Special Corporations won vigorous support from the opposition parties because the reform would impact on *amakudari*. They officially approved his proposed reforms of seven Special Corporations in November 2001.

Before his assassination by a right-wing sympathizer in Tokyo on 24 October 2002 Koki Ishii, a member of the DPJ, held a seat in the Lower House of the National Diet, serving as the chairman of the Special Committee on Disasters. He had entered politics as a member of the New Japan Party and won a seat in the Lower House in 1993. He entered the DPJ when it was founded in 1998 by Kan, Hata and Hatoyama, winning a second term in the Lower House. He began serving a third term in June 2000.

Ishii, whose concerns centered on political and administrative misconduct, contended that before structural reforms could progress, Special Corporations and other types of public corporations that use public funds and tax revenue and serve *amakudari* should be dismantled. He believed that the reluctance of the ministries to reform their corporations served to preserve their territory and to rigidify Japan's political economic system thus inhibiting the initiation of reforms.

In 1999 he published a book about *amakudari* entitled *Bureaucrat Heaven: The Bankrupting of Japan.*[8] Ishii followed this in 2001 with a book on public corporations entitled *The Parasites Gobbling Up Japan: Dismantle All Special Corporations and Public Corporations!*[9] Although his books may appear to be an effort on behalf of the DPJ to weaken the bureaucracy and loosen the ministries' ties with the LDP, his reports pose pertinent questions concerning the rapid escalation of public corporations that were established through Special Corporations and government agencies, and the employment opportunities they offer to elite bureaucrats.

Koizumi requested the ministries to review how their corporations that were in debt were spending revenue. He intended to cut funding by 50 percent but even among his cabinet members he experienced resistance. His State Minister Nobuteru Ishihara[10] recommended that only a third of the corporations should be considered for reform and Koizumi lowered his sights.

The Japan Highway Corporation, the Urban Development Corporation (UDC) and the Government Housing Loan Corporation (GHLC), Japan's leading mortgage lender, were established and managed by the former Ministry of Construction (MOC). A large portion of the fiscal stimulus packages went towards subsidizing the construction of infrastructure such as roads, bridges, dams and housing. In 2000, the government reported that FILP had loaned these corporations ¥2.1 trillion ($19.76 billion) and ¥10.4 trillion ($97.76 billion) respectively. However, the funds proved inadequate, and the MOF was forced to admit publicly that the corporations had gone over budget, that they, as well the Road Building Agency, needed subsidies to cover 'hidden costs', and that $61 billion of tax revenue would be tapped. FILP spending was out of control.

The disclosure was exactly what Koizumi needed to justify his pursuit of downsizing FILP, the privatization of PSS and the privatization and dismantling of Special Corporations. Ruichiro Hosokawa, a former managing editor of the *Mainichi Shimbun*, published an article in the 16 November 2001 issue of *The Japan Times* warning that Japan's economy would fail if reforms were not initiated. Emphasizing the enormity of Japan's public debt, he reminded readers that prior to 1975 the construction of infrastructure had been paid for with tax revenue generated by Japan's economic expansion. After that the government began issuing government bonds to finance infrastructure work such as roads and bridges. In 1976, the Japanese were enjoying a stable economy and tolerated government borrowing. However, despite Japan's recessive economy, both central and local governments were continuing to borrow with a nod of approval from the public. Hosokawa forecast that if the government continued to support half of its budget with deficit-covering bonds, Japan would become bankrupt.

Hosokawa urged the dissolution of the debt-ridden JH for the following reasons: (i) its president, vice-president and directors were former officials of the MOC and lacked management skills; and (ii) public corporations should not be allowed to operate at a deficit. He also contended that the JH's top executives should resign and refrain from accepting their retirement allowances. The same could be said of other public corporations.

Kazuma Tsutsumi, as the director of the Council of Labor Unions for Liaison with Special Corporations, was in the perfect position to observe how the ministries used Special Corporations. In his book *The Monster Ministries and Amakudari: White Paper on Corruption*, Tsutsumi claimed that a JH subsidiary, the Japan Highway Improvement Corporation, made contributions to an LDP association called the National Political Society.[11]

The privatization of the Japan Highway Corporation: unenthusiastic support from the LDP and the ministries

Koizumi devised a scheme that would convert a number of Special Corporations into Independent Administrative Institutions with the expectation that eventually financing from FILP and tax revenue would no longer be necessary and that there would be more transparency with regard to accounting methods.

Koizumi focused his initial efforts on the dissolution of the former MOC's debt-ridden JH, which Ishii labeled as 'the world's largest general contractor'. The JH had over sixty subsidiaries with such names as Sendai Highway Service, Niigata Highway Service, Hokkaido Highway Service, Osaka Highway Service, Highway Service Research, Sapporo Engineer, Osaka Highway Engineer, Sapporo Engineer, New Japan Highway Patrol, Japan High Car, Highway Kobe, Highway Toll System and Osaka Media Ports, whose expenditures in 1998 were the highest at ¥20 billion ($60 million). In theory, the JH repaid loans from FILP with revenue collected from highway tolls but the tolls were not nearly enough to cover construction and maintenance costs. In Japan the highway and bridge tolls can be more expensive than bus and ferry fares and many commuters opt to use public transport.

Originally, Koizumi wanted to merge the JH with three other corporations – the Hanshin Expressway Corporation, the Metropolitan Expressway Corporation and the Shikoku-Honshu Bridge Authority, which carried massive debts. In total, the accumulated debt was $488 billion. He then wanted to privatize the single entity by 2005 and have it repay the outstanding loans within forty-five years. He also wanted to cut government investment in future road construction by 40 percent because costs had ballooned to ¥300 billion ($2.46 billion) annually.

The plans were admirable but the implementation of them was hindered by the vested interests of LDP law-makers, who relied heavily on contributions from their constituencies, who depended on public works projects for contracts and employment, and the bureaucrats, who relied upon post-retirement positions in construction-related businesses. They demanded that the debts of the Shikoku-Honshu Bridge Authority, which could not repay the FILP loans, be separated from the other corporations, and that those prefectures where the bridges were located share the burden of the repayment with central government. Koizumi's administration was pressured to produce a watered-down version of the original package that Koizumi had hoped to get through the National Diet.

The diluted version only focused on the completion of a 9,342 km expressway, courtesy of a proposal from the MLIT. The cost was estimated to be ¥26.6 trillion ($214 billion). The repayment of the debt was doubtful. Two of the five-member advisory panel, who Koizumi appointed to review the reform and who had prioritized repayment of the debt, resigned in protest on 23 December 2003, accusing the prime minister of failing to adhere to his objectives. They stated that the new version of the bill contradicted the purposes of the original bill, which had favored the repayment of the debt over the construction. Also, the revised bill would permit the continuation of state involvement in wasteful road construction and delay debt repayment.

The process of the privatization of the JH was turbulent. The last president of the JH was Haruhiko Fujii. He was dismissed by the land minister Nobuteru Ishihara after a much-publicized heated confrontation with Ishihara who accused him of not cooperating in the process of the privatization. Fujii had taken the post in 2000 after retiring as vice-minister of the former MOC. During the period he was in both offices, Fujii was popular among LDP politicians because he expanded highway networks considerably and because of his close relationship with road construction firms.

On 14 April 2004, Ishihara replaced Fujii with his LDP colleague, Takeshi Kondo, who was a member of the Upper House. Kondo had previously been an executive of Itochu Corporation, a major trading firm. He retired in April 2006 and was succeeded by Hinori Yano, who was a former CEO of Toshiba Europe and a director of the Japan Federation of Economic Organizations.

The four public corporations were privatized on 1 October 2005. But instead of one entity there are three: East Nippon Expressway Co. Ltd. (NEXO East Japan), Central Nippon Expressway Co. Ltd. (NEXCO Central Japan) and West Nippon Expressway Co. Ltd. (NEXCO West Japan). The MLIT continued to manage the highway networks.

When Hiroshi Kato was interviewed again in June 2001 by *The Japan Times* he claimed that in the 1980s LDP officials were horrified when he recommended that the tax revenue collected from gasoline and vehicle taxes, which was targeted exclusively for road construction, also be used to finance the Japan National Railroads' enormous debts. One of the biggest financiers of the LDP campaign coffers were the construction companies.[12] The construction of public works projects is essential to local employment during economic downturns. In many rural areas, construction accounts for 20 percent of the workforce.

Special Corporations: consequences of *amakudari*

Since the JH distributed contracts to the construction companies, bid-rigging involving large construction companies was commonplace. On 29 September 2005, the Japan Fair Trade Commission (JFTC) ordered Japanese steel bridge-builders to stop bid-rigging for contracts from government and from the JH. The JFTC alleged that twenty former officials in the JH, including former Vice-President Michio Uchida and a former executive board member Tsuneo Kaneko, had received jobs in forty-five companies due to their involvement in bid-rigging. Among the forty-five firms named were Mitsubishi Heavy Industries Co., Ishikawa-Harima Heavy Industries and Kawasaki Heavy Industries Ltd. The JFTC announced that the contracts procured through illegal bid-rigging were worth approximately ¥260 billion ($2.35 billion). Former JH officials who were employed in the bridge construction companies had accessed unpublished information in the JH regarding toll road bridge construction projects.

Uchida and Kaneko, along with officials from twenty-six corporations were indicted on 8 December 2005 for bid-rigging. Some of the officials had been officers in the JH before their employment in the firms. Besides the companies listed above, other companies accused of bid-rigging included Mitsui Engineering and Shipbuilding Co., Sumitomo

Heavy Industries Ltd., Hitachi Zosen Corporation and Nippon Steel Corporation.

Uchida and Kaneko were charged with instructing subordinates to carve up contracts for an elevated highway bridge in Shizuoka Prefecture in May 2004. The contracts inflated the JH's costs by ¥47.8 million ($445,000). Uchida was arrested on 25 July 2005 and later fired from his position as vice-president of the JH on 22 August. Kaneko was arrested on 1 August.

On 17 December 2005, they entered a plea of not guilty to the charges. However, the private corporations pleaded guilty. On 7 December 2007, Kaneko was found guilty of violating the Anti-Monopoly Law and sentenced to a two-year prison term, although it was suspended for three years.

The liquidation of a failed Special Corporation: Government Housing and Loan Guarantee Corporation

Koizumi resolved to liquidate the GHLC which was heavily in debt. The former MOC established the GHLC as a Special Corporation in 1950 to provide finance for the rebuilding of homes devastated by the Second World War. It was the biggest home lender, financing 30 percent of all houses built since the war. Retail banks were unable to compete with the GHLC because of asset liability risks. In 2001, the GHLC provided 27 percent of all new loans. In 2002, 40 percent of all mortgage debt was from GHLC loans. The repayment period was up to thirty-five years for long-term, fixed-interest rate loans. The funding was mainly through FILP.

The reform of FILP would make the provision of long-term, fixed-rate loans by the GHLC difficult. The GHLC was converted to an IAI on 1 April 2007 and christened the Japan Housing Finance Agency (JHF).[13] The corporation does not engage in direct housing loans to the general public. Its main business is securitizing mortgage debt to enable private financial institutions to create a steady supply of long-term fixed-rate mortgage loans. In other words, asset liability management (ALM)[14] is passed on to the market by securitization, reducing the ALM risk for the JHF. However, the credit risk in securitized loans is undertaken by the JHF. The IAI is 100 percent government funded and managed by the minister of MOF and the minister of MLIT. So far, there have been no signs of success in the reorganization of the GHLC because of the soft real estate market.

The reform of the GHLC appears not to have stopped the misconduct of officials posted in the JHF. On 18 May 2011 Satoru Kuze, the agency's former sales promotion chief, was arrested for allegedly accepting ¥2 million ($24,420) between 2007 and 2008 in exchange for confidential information about the JHF from the former chairman and a former executive of a private home financing agency. The company wanted to become a sales agent for JHF's housing loans. While the company admitted the allegations, Kuze claimed that he had merely borrowed the money.

The image of reform: Urban Development Corporation

Another Special Corporation targeted for reform was the Urban Development Corporation (UDC), which the former MOC established in 1955 to engage in the supply and maintenance of housing. Funded by FILP, it had constructed 1.5 million homes by 1999. The corporation was reorganized into an IAI in July 2004 and renamed the Urban Renaissance Agency (UR).[15] Theoretically, the organization's work is no longer inclusive of urban development projects and projects providing housing. Its principal work is building public facilities to stimulate private sector investment in public works projects. The UR is 100 percent government funded and managed by the minister of MLIT.

The image of reform: Japan National Oil Corporation

Japan has to import 80 percent of its fossil fuel. METI oversees the energy-producing industries, among them oil. The ministry manages imports and refining through federations of oil importers. The former MITI established JNOC as a Special Corporation in 1967 to assist Japanese oil companies with exploration and drilling for oil. The corporation received funds from FILP and had 142 affiliates and branch offices overseas. The federations connected METI to the oil refiners, who distributed to retailers. The domestic companies cooperated with foreign firms, usually holding the larger share of the investment.

Koizumi wanted to privatize JNOC because in 1998 when the corporation was targeted for restructuring the president Kuni Komatsu divulged that the company carried outstanding debt of ¥154.2 billion ($1.23 billion). However, Japan's oil refiners were opposed to private companies taking over JNOC and wanted the government to continue JNOC's operations because the corporations were financially too weak to take on the risks of oil exploration.

In August 2003, JNOC declared a net loss of $1.9 billion for fiscal 2002 and an accumulated debt of ¥770.1 billion ($6.17 billion) due to the failure of its subsidiary the Japan Oil Development Co. (JODCO). In 2004 METI began to dismantle JNOC, privatizing some of its subsidiaries such as the Japan Petroleum Exploration Co. and the Indonesia Petroleum Exploration Co. (INPEX Corp.), a major upstream oil and gas exploration company that was established in 1966 of which it owned a 53.96 percent share. INPEX had expanded operations internationally and was popular among foreign investors.

METI agreed to clear JODCO's debts in order to convince INPEX to take over the money-losing JODCO. INPEX merged with Teikoku Oil in March 2006, receiving 81 percent of its shares. By 2008, JODCO was fully integrated into INPEX Holdings.

METI retained JNOC's oil and gas exploration units when it was merged with its Special Corporation the Metal and Mining Agency, which in 2004 was christened as an IAI the Japan Oil, Gas and Metals National Corporation (JOGMEC).[16] JOGMEC engages jointly with oil companies in the exploration and production of oil and natural gas.

One of the major complaints lodged against JNOC was poor management by officials who did not have the expertise to direct oil exploration and production, pointing to METI officials who took temporary positions for two years and forged relationships with domestic and foreign oil companies, which led to permanent post-retirement upper management positions in those companies. Traditionally, a retired administrative vice-minister from the ministry served as president in JNOC while retired MOF officials filled other top management positions such as vice-president or the director of finance. Before his resignation, Komatsu had served as president for six years. Although he had climbed to the position of MITI Administrative Vice-Minister he had no experience in energy administration. In his book Kazuma Tsutsumi lists a number of such migrations of officials from JNOC and from other MITI Special Corporations to such companies as Indonesia Oil, Japan Steel Pipe Co., Mobile, Shell and Abu Dhabi Oil.[17]

Amakudari was allowed to continue in JETRO, a Special Corporation. It was a coup for METI since JETRO's operations began receiving negative press in Japan and in the United States about its operations in 1995.

The image of reform: Japan External Trade Organization

The former MITI established JETRO as a Special Corporation in 1956 to promote Japanese small businesses to export goods to lucrative markets and to provide them with information on foreign markets. In 1958,

JETRO created the Institute of Developing Economies (IDE) as a separate public corporation but it was merged with JETRO in 1998. In 1971, yet another public organization was established – the International Economic and Trade Information Center.

JETRO opened its first overseas offices in London and in New York City in 1959, registering in the United States as a public corporation and not as an agency of a foreign government under the Foreign Agents Acts of 1938, which caused some consternation among American officials. In 1976, the US Department of Justice sued the Japan Trade Council (established in Washington, DC in 1958) for civil fraud, charging that MITI contributed 90 percent of the Council's funds through its JETRO New York office. Soon afterwards, JETRO reregistered as a foreign agent.

By 1975, JETRO was operating twenty-four trade centers and fifty-four offices in fifty-five countries, testimony to the fact that not only had Japan become a major player in world markets, but also that MITI was putting down roots overseas. Besides the JETRO offices, MITI officers were loaned to Japanese consulates, embassies and Japan Chamber of Commerce offices for periods of two to three years.

By the early 1980s, Japan, whose economy was export-driven, was showing a marked trade surplus with its leading partner, namely the United States, which was in recession. There was significant pressure from the US government to deregulate domestic markets and raise import quotas for such goods as agriculture products, electronics and motor vehicles. Realistically, JETRO's role as a promoter of Japanese exports was no longer as relevant to Japanese businesses as it had been in the 1960s and 1970s. Also, JETRO's role as a surveyor of foreign markets and a collector of economic and political data had, in part, become extraneous because the research was also being conducted by large Japanese multinationals, research institutes and by MITI officers posted in embassies and Japan Chamber of Commerce offices.

Manipulation of operations to maintain JETRO

JETRO is a prime example of how the ministries can redesign their corporations in order to preserve them. MITI intended to continue operating a corporation that had effectively resulted in creating more territory for its officials and, thus, more opportunities for *amakudari*. As a gesture of compliance with US demands, MITI began the process of re-orchestrating JETRO's functions so that the organization would serve as a promoter of foreign imports and foreign investment.

In 1983 JETRO set up a taskforce to look at import promotion. In 1984, import promotion activities such as trade fairs were held in Nagoya, Yokohama and Kitakyushu. In the same year, a second taskforce was set up to promote international economic cooperation for industry. In 1989, MITI completed the conversion of JETRO export promoter to JETRO import promoter, with the establishment of yet another organization, the Institute for International Trade and Investment. JETRO's foreign offices increased to eighty in fifty-seven countries (the number has since decreased to seventy-three offices in fifty-five countries) while domestic offices throughout Japan increased to thirty-eight.

Although JETRO offered free to the public government-generated data on Japan's economy, trade statistics and information on Japanese culture and lifestyles, JETRO's primary image was that of a Japanese government-sponsored organization providing support to business owners who want to export goods and services to Japan. JETRO began publishing a series of market reports for products that found Japanese consumer acceptance such as foods, alcoholic beverages, clothing, sports equipment, cosmetics, electronic equipment, jewelry and organic produce. More than a hundred reports had been published by the end of the 1990s. JETRO's *Success Surveys* told of businesses that had entered the Japanese market successfully. There were trade directories of Japanese companies who were interested in ventures with foreign businesses. The materials gave a positive depiction of easy penetration of Japanese markets for owners of small businesses. If they were willing to adapt their products to meet Japanese consumer preferences, and understand the Japanese way of doing business they could experience success like the companies described in the JETRO publications.

JETRO also offered free consulting services at its offices for prospective exporters as well as a number of other support services in Japan to foreign small and medium-size business owners. Trade fairs focused on specific industries where businesses could display products and meet potential buyers. JETRO support centers were located in many of the Foreign Access Zones (FAZ) providing information to foreign businesses about Japanese markets and doing business within FAZ. In 1993, business support centers (BSCs) were opened in Tokyo, Yokohama, Kobe, Nagoya and Fukuoka to assist foreign small-business owners during their visits to Japan to find buyers for their products. The BSCs[18] offered exhibition space, temporary facilities and consulting free of charge.

The Foreign Investment in Japan Development Corporation (FIND) was established as a private corporation in 1993, promoting itself as 'Your Foothold in the Japanese Market'. Its services included the contribution of funds to foreign businesses that wanted to operate in Japan. The corporation claimed that capital had been provided by government and by powerful businesses and industrial associations. Among its thirty-one stockholders were METI industrial associations such as the Electronics Industries Association of Japan (EIAJ), a METI subsidiary, and the Industrial Structure Improvement Fund, which owned a major stake in FIND. The majority of private corporations that were investors were in industries that METI regulated. The companies included Toyota Motors Corp., Mitsubishi Chemical Corp., Nissho Iwai Corp., Kobe Steel Ltd. and Nissan Motors Corp.

In his book *Troubled Times* published in 1999 Dr. Edward J. Lincoln stated that it had been anticipated that FIND would assist in a concrete way with business investment, but the corporation was criticized because it did little more than propose joint ventures with the Japanese firms who were members of FIND. It also charged a fee for the introductions. Lincoln argued that FIND was a government corporation and was therefore not free to give advice on mergers and acquisitions. He concluded: 'Foreign firms were less in need of advice or introductions to potential businesses than in a dismantling of the real obstacles to acquisitions.'[19]

Lincoln was referring to government's stiff regulations regarding foreign acquisition of Japanese firms, indeed, regulations that METI helped to create. FIND was closed down on 31 March 2002 to be replaced by INVEST JAPAN when JETRO was converted to an IAI on 1 October 2003.

JETRO's import promotion programs and literature were packaged to give the corporation a glossy professional image of a Japanese government funded agency that was earnest in its efforts to help small businesses enter the Japanese markets. However, during the 1990s this image became tarnished. Complaints from dissatisfied business owners reported disappointing results from JETRO's services such as the BSCs because they were not centrally located, impacting on the number of visitors and potential buyers.

The disguise: convincing the Japanese media

The 9 January 1995 Internet edition of *Asahi Shimbun*'s 'Opinion Page' noted that when the Murayama administration conducted hearings on the restructuring of Special Corporations, the former MITI was reluctant to participate and wanted to know if the restructuring concerned the number of corporations or if the discussion was related to the financing of the corporations. The article also claimed that the ministries were changing the objectives of their corporations by contriving new roles. The paper called this 'skill at disguising' (*henshin no gijutsu*), pointing to JETRO as an example of a Special Corporation that had been established in 1956 for the purpose of promoting Japanese exports: 'Now when you phone JETRO headquarters the receptionist answers: "JETRO, import promoter".'

Japanese media focus on JETRO's operations continued. On 13 June 1995 the front page of the *Sankei Shimbun* published an article by its Washington correspondent Yoshihisa Komori titled 'Is JETRO Running Wild?' ('*Jetero Boso?*'). Next to the story was a photograph of JETRO's Tokyo headquarters. Komori pointed his finger to the editorial in the April edition of a monthly newsletter *Inside/Outside Japan* published by JETRO New York and sent unsolicited to opinion leaders in business, government and academia. In his editorial, Nobuya Noguchi, the president of JETRO New York and a MITI official, wrote positively about former US Secretary of Defense Robert McNamara's book *In Retrospect: The Tragedy of Lessons of Vietnam*.

According to Komori, one reader, William Triplet II, an aide to Republican Senator Robert Bennet, who was a member of the Senate Committee of Foreign Relations and a recipient of the newsletter, took exception to the article, protesting that the president of an organization that was established to promote foreign trade and economic cooperation should not be involved in commenting on political issues.

Komori claimed that there was opposition in the National Diet to the continuation of JETRO because (i) it no longer served its original function as a trade organization; (ii) JETRO was an underground MITI (*kakure Tsuusansho*); and (iii) JETRO had in effect become the 'Number Two Ministry of Foreign Affairs' (*dai-ni Gaimusho*). By alleging that JETRO had become a second Ministry of Foreign Affairs (MOFA), *Sankei*, a major daily known for its conservative stance, was implying that MITI was using JETRO to wrest territory from MOFA.

The monthly political magazine *Sentaku* followed with an article in its July 1995 issue. The story carried a photograph of *Inside/Outside Japan*. The article claimed that the reason Triplet was annoyed with Noguchi's article was because JETRO's activities in the United States were a source of irritation to the CIA and FBI. *Sentaku* asserted that the CIA and FBI were watching closely the activities of the directors of industrial research in JETRO New York when they visited other JETRO offices in the United States. The agencies regarded the MITI officers as CIA-type agents from Japan and since the representatives could not be classified as either foreign diplomats or scholars, their status was ambiguous. Also, there was suspicion among members in Congress that the officers engaged in industrial espionage.

The disguise: convincing the American media

The 16 June 1997 issue of the *U.S. News and World Report* probed JETRO's authenticity as an import promoter. An article by senior writer William J. Holstein[20] entitled 'With Friends Like These'[21] questioned JETRO's function in the United States. In his article, Holstein described JETRO as 'a uniquely flexible organization that defies American definition'. He contended that JETRO America did not serve to promote imports into Japan, but rather was a sophisticated commercial-intelligence-gathering agency. He suggested that the promotional materials served to disguise the true reasons for JETRO's presence. Dr. Edward Lincoln, who was an advisor to former ambassador to Japan Walter Mondale, told Holstein: 'At best the Japanese are being disingenuous when they say that JETRO's primary job is promoting American exports.' Lincoln stated that JETRO's 'core mission' was to collect American technology and political intelligence.[22]

In his article Holstein stated that the products invented by small businesses would be appropriated by Japanese companies to whom they had been introduced by JETRO. He presented as an example an entrepreneur in Florida whose electronic device JETRO claimed in its publication *Success in the Making* to have marketed successfully in Japan. Holstein contended that JETRO's support led to the inventor's loss of control over his invention to the automobile manufacturer, Toyota Corp. After Toyota had agreed to use the device for a few days, the Japanese distributor advised the inventor that it would be best to sell because Toyota was a giant manufacturer and very influential.[23]

The disguise: convincing Koizumi

The 2002 JETRO website stated: 'JETRO made a 180-degree reversal and began promoting imports entering Japan, a primary mission to this day.'

Members of Koizumi's cabinet also questioned JETRO's role. Cabinet State Minister of Administration and Regulatory Reforms Nobuteru Ishihara told reporters at a press conference on 22 November 2002 that the continuation of Special Corporations was problematic and that he hoped that the members in the Diet would cooperate. He singled out

JETRO as an example of a corporation that was no longer serving the function it was originally established to serve and that work was being contrived to justify its budget and to continue operations. He stated that both he and the Minister of Economy Trade and Industry were surprised to learn that the corporation issued a pamphlet advertising import promotion and not export promotion.

On 1 October 2003 JETRO was converted to an IAI. It was business as usual with METI controlling the corporation's budget. *The Japan Times* reported on 25 October 2009 that the chairman, a retired METI official, received an annual salary of approximately $273,000. The average income of other retired ministry officials who landed top posts in IAIs was $220,000. All of the services listed above are currently offered. The 2011 website states:

> The Japan External Trade Organization (JETRO) is a government-related organization that works to promote mutual trade and investment between Japan and the rest of the world. Originally established in 1958 to promote Japanese exports abroad, JETRO's core focus in the 21st century has shifted toward promoting foreign direct investment into Japan and helping small–medium size Japanese firms maximize their global export potential.

JETRO's 'core focus': sowing the seeds of the ministry

The thirty-six JETRO domestic offices and seventy-three overseas offices in fifty-five countries offer METI ample opportunity to post officials in the prefectures and abroad. JETRO offices also serve to extend METI's presence. METI opened its seventh JETRO office in China in Wuhan on 17 July 2011 to support Japanese business expansion in Henan, Hubei, Hunan and Jianxi provinces where rapid economic development was anticipated.

METI created another public corporation 2003 by using the Institute of Developing Economies (IDE) that had been previously merged with JETRO in 1998 and joined it with JETRO to form the Institute of Developing Economies–Japan External Trade Organization (IDE–JETRO). IDE's website states that the institute 'conducts research on economic, political and societal issues in developing economies to support Japan's expansion of harmonious trade and investment and the provision of international economic cooperation focused on developing economies'.

METI's policies are still influenced by the former MITI's protectionist policies. On the other hand, the ministry ostensibly promotes inward investment to Japan and uses JETRO for these activities. So far, the promotion has proven ineffective. In 2007, the United Nations Conference on Trade and Development (UNCTAD) inward FDI performance index ranked Japan 137th among 141 countries, which constituted 2.5 percent of the GDP. During the last two years, Japan's inward investment has increased to approximately 3.5 percent but it is still among the lowest of the UNCTAD countries. Peter Mandelson, the former European Union Trade Commissioner, complained in a speech given at the EU–Japan Center for Industrial Cooperation on 21 April 2008 that Japan was 'the most closed investment market in the developed world'. Ironically, the event was sponsored by METI and JETRO. Mandelson suggested that Japan, whilst taking advantage of the openness of foreign markets was creating barriers to foreign investment. He cited figures showing that only 3 percent of Europe's total $300 million outward investment was invested in Japan, comparing it with Japan's outward investment. 'For every dollar Japan invested in the UK and the Netherlands alone, European companies were able to invest a net total of only 3 percent in Japan.'

Holding on to a good thing

The secretariat of the cabinet's Headquarters for Administrative Reform on 8 November 2007 released documents to a government panel of experts on streamlining public corporations. Originally, fifty-seven corporations were set up to take over part of the operations of the ministries and agencies but this number had increased to 101, a similar pattern to that which evolved as the ministries were establishing their Special Corporations. The report showed that 40 out of 101 IAIs awarded contracts to their subsidiaries with more than 90 percent of the contracts completed without competitive bidding.

The documents also revealed that the 101 IAIs altogether had 236 affiliates. Companies where one-third or more of their posts were occupied by former directors and senior officials of the IAIs were among the affiliates. Also, 230 officials at the 101 corporations had assumed directors' posts at affiliates in fiscal 2005 through *amakudari*.

In addition to JOGMEC, METI manages three other IAIs that are directly related to the energy sector:

1. The New Energy and Industrial Technology Development Organization (NEDO)[24] was established by the former MITI as a Special Corporation in 1980 for the promotion and funding of projects

related to renewable energy and the development of industrial technologies. With a budget of about $2.9 billion (FY 2009) NEDO employs 1,000 staff and engages in joint projects overseas with offices located in Silicon Valley, Washington, DC, Paris, Beijing, Bangkok and New Delhi.

2. The Japan Nuclear Energy Safety Organization (JNES).[25]
3. The National Institute of Advanced Industrial Science and Technology (AIST)[26] was founded in 2001 as an IAI as an amalgamation of fifteen research institutes that were managed by the former MITI. Its predecessor was established in 1882. The new AIST is the largest Japanese government-supported research institute in Japan with forty autonomous research institutes employing 2,400 researchers and 700 administrative staff.

Despite the name-change to Independent Administrative Institutions the Japanese still refer to the ministries' public corporations as Special Corporations (*tokushuhoujin*). In 2003, Carpenter stated that 'Special Corporations are illustrative of the basic nature of Japan's political economy. The ministries' determination to maintain territory and thus protect vested interests can be seen in the continued operations of Special Corporations despite Koizumi's plans to dissolve them.'[27]

2

Amakudari in the Ministries' IAIs, Public Corporations, Research Institutes and Affiliated Agencies: the Insidious Side

In his book *Troubled Times* Lincoln (1999), pointed to a key reason for Japan's closed markets:

> The *amakudari* system provides substantial reason to be skeptical of the extent of deregulation and the unilateral market opening in Japan because of the manner in which this practice establishes a broad web of personal ties between government and Japanese firms.[1]

Foreign firms also employ ministry officials to sit on their boards. IBM is a good example of this practice. MITI's post-war industrial policy was the development of energy-intensive industries such as aluminum and petrochemicals. When these industries suffered structural damage as a consequence of the first oil shock in December 1973 MITI was compelled to plan policies that created new industries that could compete in international markets. Promoting the development of information technologies was vital to MITI's objectives. Fujitsu Computers was nurtured by MITI through heavy subsidization and protection from foreign competition during the 1970s and 1980s.

When International Business Machines (IBM) entered Japan in 1976 in the information business service sector it found that the market was dominated by Japanese IT companies, led by Fujitsu. It also experienced competition in international markets because the subsidized Fujitsu could undercut prices through 'dumping'.

In order to compete with Japanese companies IBM imitated other foreign multinationals that wanted a piece of Japan's protected market. According to Calder (1989), in the 1980s IBM hired officials from MOF, the BOJ and the Science and Technology Agency and, including three retired MITI officials, the company had more retired senior bureaucrats on its board than any of the Japanese computer manufacturers.[2] The strategy succeeded in bringing IBM lucrative contracts.[3]

Amakudari is the reemployment of retired or incumbent civil servants in private and public sector corporations or in government-affiliated entities. The literal translation of the term is 'descent from heaven', which refers to high-ranking ministry officials whom the Japanese considered to be the elite among government officials. *Amakudari* and interpersonal networks between the civil service and the private sector can be traced to the Meiji period (1868–1912) when the bureaucracy collaborated closely with large family-owned conglomerates to develop feudal Japan into an industrialized country in the same league as the Western powers by 1928. The collaboration intensified during the Second World War when the government started to strictly regulate the economy for the war effort. Business owners would employ bureaucrats in order to determine future government directives and also to lobby interests.

Amakudari became a formalized institutional arrangement as a part of the post-war civil service system. The system benefits officials who retire earlier than staff in the private sector, usually between the ages of 55 and 60, stepping aside for the younger officials who are intent on promotion. The ministries want to provide retiring officials with a source of income that supplements their pensions, which can be lower than for employees in the private sector. Ostensibly the arrangement acts as an incentive for university graduates who would otherwise seek employment in the private sector rather than the civil service, where salaries can be lower.

However, the system also serves the interests of the ministries and their former colleagues throughout the public and private sectors and enhances the bureaucracy's power to control the implementation of economic and industrial policies. *Amakudari* expedites ministerial guidance over the businesses that are within a ministry's administrative jurisdiction. The relationship between bureaucrats and former colleagues who have migrated to businesses in a sector under their ministry's administrative jurisdiction automatically tightens the ministry's grip on that sector. Originally, the elite bureaucrats were the main beneficiaries of *amakudari* but, gradually, officers in middle management as well as non-career officers were also included in the system.

The system has served as a mutual back-scratching mechanism because the companies that hire bureaucrats have a direct pipeline to the ministries and accept officials with the anticipation that they will be treated favorably and receive useful information and subsidies, contracts for public works, as well as swift approval of such applications as licenses and patents. Companies may be reticent to employ officials but will acquiesce to pressure from the ministries because of concerns that they will not be able to compete with businesses who have hired officials.

No reforms in sight

In January 2006, a government report announced that in December 2005, 34 percent of senior bureaucrats who had retired from their ministries in August had taken jobs in ministry-affiliated agencies. The following April, data were released showing that 70 percent of 1,968 bureaucrats had been assisted by their ministries and agencies to land jobs for the period 2004–6 in the sectors under the administrative jurisdiction of their organization. In response, the Ministry of Internal Affairs and Communications (MIC) took steps in August 2006 to implement new rules regarding post-retirement positions for national ministry officials.

In March 2007, a panel led by former LDP Prime Minister Shinzo Abe submitted a plan that would prohibit the ministries finding post-retirement positions for their officials and would create a new body that would centralize job-hunting for all of the ministries. Nevertheless, it was estimated that it would take at least three years to begin operations after its establishment. Abe's comments at the panel discussion summed up the struggle to curtail *amakudari* practices: 'I have said that we have to root out the ministries' somewhat forcible job-hunting tactics that are backed by their budgetary power and authority.'

In April, Abe's cabinet approved a bill that would end *amakudari*. The prime minister touted the bill as the biggest reform of the administrative system in sixty years. The bill was submitted to the Diet in June. Besides opposition to the bill from Abe's LDP colleagues, the opposition parties were also against it, contending that the bill did not guarantee the end of collusion between government and the private sector. Abe's coalition government, frustrated by the opposition, temporarily withdrew the bill from the Diet.

Nevertheless, the struggle to eliminate *amakudari* continued a year later in April 2008 when former Prime Minister Yasuo Fukuda's LDP–New Komeito coalition government administration announced a draft bill for further reform and regulation of the 101 IAIs. Applicants from outside the ministries would be permitted to apply for positions in the ministries' organizations but appointments would be approved by all members of the cabinet. Also, retired bureaucrats working in IAIs would be prohibited from slipping into businesses that had connections with the organizations.

Opponents of the bill were skeptical that it would eliminate *amakudari* since the cabinet systematically approved appointments of senior ministry officials to positions in IAIs.

Fukuda's successor Aso established the Center for Personnel Exchanges between Government and Private Entities in order to integrate job placement for civil servants. The Center is operated by the National Personnel Authority (NPA) in the MIC. Aso's objective is to put a stop to rampant bid-rigging for public works at the initiative of government officials, which leads to wasteful spending of tax revenue, and to prevent ministries and agencies from directly contacting interested companies and other entities to find jobs for its officials.

However, the new system merely served to promote the image of reform. Since its inauguration in August 2009 the Center placed only 172 senior bureaucrats in jobs and out of that number 81 were hired by the ministries' public corporations. In fact, the total number of retired ministry officials migrating to jobs in government-affiliated agencies and the ministries' IAIs has increased substantially since 2009. From September 2009 to October 2010, the government recorded the migration of 4,240 bureaucrats to IAIs, the ministries' agencies and the public utilities. Furthermore, the ministries are not required to submit applications to government in order to place officials in public corporations or their affiliated agencies.

The NPA actively sought upper-management positions in the private sector for retiring ministry officials prior to the establishment of the Center for Personnel Exchanges. Nevertheless, the ministries also search on behalf of their staff usually in companies and other entities in the sector they administrate using their budgets and authority to exert pressure. MOF officials will find employment in the financial industry, while METI officials find employment in manufacturing, retail, energy industries and power utilities. As an example of METI officials migrating to TEPCO, Toru Ishida, the former director-general of METI's ANRE was employed by TEPCO in January 2011.[4]

As a part of its strategy to dismantle Japan's industrial conglomerates at the end of the Second World War, the Supreme Commander of the Allied Powers carved up Japan's national electricity provider, the Japan Electricity Company, into smaller private electric power utility companies. TEPCO was established in 1951. Since there were few private companies that could deliver wholesale electricity to the private companies or which had the funds to invest in the R&D of more efficient power sources the government established the EPDC in 1952 to supply electricity wholesale to electric power utility companies throughout Japan. The corporation was managed by the former MITI. The ministry's officials received post-retirement positions in the EPDC and its subsidiaries, enabling them to establish close links with the electric utility companies and to migrate to positions in those companies.[5] When the EPDC was privatized in 2004 and renamed J-POWER the names of its subsidiaries were also changed with the exception of KEC Corporation. Besides coal and hydroelectric power stations J-POWER constructs nuclear power plants and is currently planning the construction of a reactor in Oma, Aomori Prefecture.

On 3 May 2011, Kyodo News Service reported that during the last fifty years sixty-eight former METI elite officials took top positions in Japan's twelve electricity suppliers. Five of the positions were in TEPCO. In 2011, thirteen retired METI bureaucrats were in senior positions at electric power companies through *amakudari*.[6] Although METI Minister Banri Kaida asked METI officials to refrain from accepting jobs in government-affiliated institutions or companies in METI's administrative jurisdiction, he did not have the authority to compel retired officials to leave their jobs.

Ministry officials can be placed directly upon retirement in corporations that are not directly within their ministry's administrative jurisdiction.

Manifestations of *amakudari*: IAI/public corporations

At the 13 April 2011 press briefing Chief Cabinet Secretary Yukio Edano told reporters that he did not know whether Ishida had received employment at TEPCO 'legally', referring to the National Public Service Law which stipulates that bureaucrats cannot, for a period of two years, legally move directly to positions in private companies attached to the sectors their ministries regulate. Nevertheless, they can move immediately to IAIs, other types of public entities, research institutes, industrial associations and ministry-affiliated agencies where they linger for two years on salary before going on to the private sector. The term 'retired' refers to bureaucrats who are within ten years of retiring from their ministries.

Edano requested that senior METI officials, including officials in both NISA and ANRE, temporarily abstain from taking positions in power companies 'in order not to raise the public's suspicions'. The ministries have devised creative methods to guarantee jobs and lucrative salaries for officials in IAIs and other public entities:

1. Bureaucrats can be sent by their ministries to work in IAIs, other types of public corporations, research institutes and industrial associations while they are still engaged by their ministries (and drawing

salary). This practice is known as 'on loan to another company' (*shukko*). Although the officials are still connected to their ministries they are identified as officers of the organizations where they have been transferred. And while the posts are considered to be temporary, they can develop into permanent positions. Essentially *shukko* can be the catalyst for *amakudari* in IAIs or in other public entities.

A survey released by the Lower House in March 2007 reported the number of IAIs (including subsidiaries) maintained by the ministries and the number of 'retired' bureaucrats, including officers on loan to these entities:

(i) The Ministry of Land, Infrastructure and Transport: 834 entities, 6,386 bureaucrats.
(ii) Ministry of Health, Labor and Welfare: 709 entities, 4,007 former officials.
(iii) Defense Ministry: 207 entities, 3,917 former officials.
(iv) Ministry of Education, Culture, Sports, Science and Technology: 934 entities, 3,007 former ministry officials.

2. The ministries are not required to apply to the NPA in order to 'loan' an official to an organization. Officials who are 'on loan' to an organization can then migrate to the private sector without waiting out the two-year period of grace.
3. Bureaucrats can remain attached in some way to IAIs/public entities through their subsidiaries (as advisors or members of boards) while working for private businesses, receiving salaries simultaneously from both IAIs and the private businesses.[7]
4. Positions in IAIs may be given to elite bureaucrats who have not reached retirement age, but are considered to be nearing the end of their careers. The 'gift' is actually a signal to officials that they will not be promoted much higher in their ministries, but that their loyalty is appreciated.[8]
5. The lucrative salaries that top management in IAIs receive make the positions enticing to retiring officials and to officials who have left their ministries to work in the private sector. On 24 October 2009, Kyodo News Service reported that ninety-eight IAIs hired retired bureaucrats for top management positions. Forty of the entities hired officials who had retired from the ministries which managed their respective IAIs. Eleven were supervised by METI and seven of them, including JETRO, were headed by former METI bureaucrats. Of the seven, six were officials who had left the ministry to work as executives and advisors at major private companies. The average annual salary of twenty-nine chiefs for FY 2009 was $226,470.

6. The ministries operate research institutes where officials are 'loaned' for a two-year period of grace before migrating to the private sector. METI's IAI research institutes include the Institute of Trade and Industry (REITI), the National Institute of Advanced Industrial Science and Technology (AIST) and the New Energy Research Organization (NEDO).

Energy-related research institutes managed by METI include the Institute of Energy Economics (IEE), the Institute of Applied Energy (IAE) and the Research Center for Hydrogen Industrial Use and Storage (HYDROGENIUS).

Commonly, administrative vice-ministers gain permanent positions in these organizations, some of which have been subsidized with funds from public corporations managed by the ministries. Officials from other ministries and other government agencies also are 'loaned' to the institutes. The ministries' research institutes and industrial organizations can also be mechanisms that link officials to private corporations.

Issues arising in research institutes

The Industrial Policy Research Institute (IPRI) was supported until 1993 by funds generated by the former MITI's Special Corporation, the Japan Corporation for the Promotion of Bicycle Racing (Japan Keiren). The taxes from bicycle racing provide substantial revenue. The unused funds for the research were given to the ministry's Secretariat Planning Office. After the funding system stopped the remaining funds were not returned to the IPRI but kept by the Planning Office in thirty-eight bank accounts.[9]

Japan Keiren continued to subsidize events and conferences. For example, Japan Keiren sponsored the 6th Asian Forum for Information and Technology (AFIT) held in 2008 which was also sponsored by METI, the Japan Electronics and Technology Industries Association (JETIA), which is a METI industrial association, Hitachi Ltd., Fujitsu Ltd. and NEC Corporation.

The Fujitsu Research Institute (FRI) was established in 1986 as an organization involved in information technology research. The institute is said to be a private think tank but it also engages with METI in numerous joint projects and the senior managing directors have been METI officials. The following example is indicative of how interpersonal networks in research institutes can encourage a mutuality of vested interests.

The former governor of the Bank of Japan, Toshihiko Fukui (March 2003–March 2008) was implicated in a scandal in 1997 concerning the entertainment of MOF and BOJ officials.[10] Fukui, who entered the BOJ in 1958 after graduating with a law degree from Tokyo University, was the deputy-governor of the BOJ at the time and, although he was not indicted, he resigned during the investigation and moved to the FRI as chairman in 1998. While he was at the FRI, Fukui invested ¥10 million ($90,000) in a stock fund operated by a former elite METI official. Yoshiaki Murakami was 46 when he left the ministry in 1999 to open M&A Consultants (also known as the Murakami Fund). One of his main backers was a close friend, Yoshihiko Miyauchi, the chairman of ORIX Financial Services, a giant financial services company.[11]

Murakami also approached Fukui and six other Fujitsu staff to purchase shares in the asset management firm. By 2002, the fund was said to manage $470 million in assets, enough to buy large shareholdings of listed companies for corporate takeovers. Murakami was popularly known as an aggressive and successful entrepreneur in the same style as Livedoor's Takafumi Horie, but his relationship with Horie brought about his downfall when he was arrested on 5 June 2006 on charges of insider trading involving Horie.[12]

The investigation by public prosecutors revealed that Fukui had not sold his shares in the fund when he assumed the post of governor of the BOJ in March 2003 (former BOJ staff can resume employment in the BOJ). Although Fukui disclosed personal assets and stocks internally, there is no law that stipulates that BOJ governors must publicly disclose their personal assets. Fukui had not transferred the shares to a third party nor had he put the shares in a trust, thus enabling him to sell the shares at leisure. On 21 June, admitting that his investment had earned ¥22.31 million ($201,000) in seven years (as of December 2005), Fukui apologized for what could be construed as a conflict of interests.

It was further revealed that two months before he assumed his post, Fukui had purchased 10,000 shares in Mitsui O.S.K. Lines; 5,000 shares in Fujitsu Ltd.; 5,000 shares in Kikkoman Corp.; 5,000 shares in Mitsui Fudosan Co.; and 10,000 shares in Nippon Steel Corp. The market value of the total shares was ¥13.31 million ($115,000) in January 2003, two months before Fukui became governor. At the end of May 2006, the shares were worth ¥34.76 million ($299,000). It was also disclosed that Fukui had served as an outside director at Fujitsu Ltd., Mitsui O.S.K. Lines Ltd. and Kikkoman Corp.

According to a public opinion poll taken by the *Asahi News* a few days after Fukui's public apology, 62 percent of the Japanese polled wanted Fukui to resign, even though he had not violated any law. Nevertheless, former Prime Minister Koizumi supported Fukui's continuation as governor.

Fukui pledged to liquidate all of his shares and donate the profits to charity. He also promised to retain only six months of his annual salary.

A consequence of *amakudari* in IAIs: bid-rigging

Even though an anti-collusion bill was implemented in 2003 to prevent bureaucrats from positively influencing the outcome of bids from corporations, collusion continued to be a common practice.

The DPJ's 2009 manifesto called for the 'elimination of *amakudari* and public officials who are involved in bid rigging'. However, merely changing the name of Special Corporations to Independent Administrative Institutions and merging some of the bankrupt ones with solvent ones did little to stop wasteful spending of tax revenue or the migration of bureaucrats to IAIs.

The *amakudari* system is a breeding ground for bid-rigging activities and collusion among bureaucrats and businessmen. The ministries' IAIs grant government contracts for public works projects to private industries and are the vehicles that connect the ministries to these industries. Former bureaucrats can choose the recipients of the lucrative contracts that are funded by tax revenue and, in return, receive employment in the corporations to whom they have awarded the contracts. A Lower House survey showed that in 2006 alone, of the $17.1 billion in disbursement for state projects, 98 percent was contracted to those entities without bidding procedures.

In December 2006, reacting to reports of bid-rigging by local government officials together with bureaucrats in IAIs, a revised law was enacted by government that would bring stiff punishment to officials in these corporations who were convicted of involvement in bid-rigging. The revised law included officials in the expressway corporations that were the result of the privatization of the JH.

The report issued by the Cabinet Headquarters for Administrative Reform released on 8 November 2007 showed that 40 out of the 101 IAIs awarded contracts to their subsidiaries with more than 90 percent of the contracts completed without competitive bidding.

An incident that clearly illustrates how *amakudari* connects IAIs with the private sector in encouraging bid-rigging occurred in April 2007 when five corporations were raided for rigging projects ordered by the Japan Green Resources Agency. The agency was known until October 2003 as the Japan Green Resources Corporation, which was a Special Corporation that was established by the Ministry of Agriculture, Forestry and Fisheries.

The five entities, as of 2005, had hired 256 former bureaucrats through *amakudari*: forty-one of the forty-four executive posts were former officials of the Forestry Agency. The Japan Forestry Foundation, administrated by the Forestry Agency, had accepted, as of April 2005, 138 bureaucrats, including the chairman who was the head of the Forestry Agency. The Japan Forest Engineering Consultants hired forty-five bureaucrats, the Japan Forest Technology Association hired forty and the Japan Federation of Forestry Civil Engineering Research Institute employed twenty-four.

According to the investigation by the Japan Fair Trade Commission (JFTC), the Forestry Agency officials were instrumental in introducing staff to firms in the private sector. The five entities were among the contractors of Japan Green Resources which were raided by the JFTC. In May 2007, Agriculture Minister Matsuoka Toshikatsu who was under investigation for allegedly receiving political donations from Japan Green Resources contractors committed suicide. A few days later, Yamazaki Shinichi, the former head of the Agency whose home was searched by Tokyo District Special Prosecutors office, also committed suicide.

A consequence of *amakudari* in an IAI: incompetent management of the Social Insurance Agency

The social security system comprises pension and health insurance. The system that was established in 1961 was reorganized in 1986 into two tiers: the Basic Pension with fixed benefits and a second tier of remuneration-based benefits (to replace the Employees' Pension Insurance) and corporate pension plans offering more options. The Basic Pension Plan covers part-time workers but the state is the insurer.

The mutual aid pensions were administrated by three Special Corporations under the supervision of the national ministries. Due to huge deficits, the three corporations were consolidated in 1997 into one Special Corporation, the Social Insurance Agency (SIA), and managed by the former Ministry of Health and Welfare. It is one of the world's largest with over 70 million members and ¥120,000 billion ($1.48 billion) in reserves. In 1997, the SIA began incorporating three different databases into one database system.

The Japanese have developed an abiding mistrust of their social security system for a number of reasons:

1. The low fertility rate is causing major concern. Twenty-five percent of the population is over 65 and the longevity rate is the highest in the world. Japan's population reached a peak of 127.77 million

in 2004. In 2005, Japan began to lose population and the National Institute for Population and Social Security Research predicted that the population would decrease to 90 million by 2055.

2. In 2009, the number of contributors dropped 0.9 percent, or 68.74 million. The proportion of the population aged 65 and older was 29 million in 2010, or 23 percent of the population. It is estimated to rise to 27.8 percent in 2020 while the population of those aged 15 to 64 will fall to 60 percent.[13]
3. Currently, the ratio of working-age population to the aging population is 3:1 (it was 10.1 in 1960). With little immigration and low birth-rate, the ratio is forecast to be 1:1 in 2055.
4. Pension and health-care payouts are swelling as the number of elderly people increases. During the last decade pension payouts have spiraled more than 30 percent. In 2009 the MHLW, which manages the public pension scheme, reported that the number of pension recipients rose by 3.1 percent, or 37 million people. Pension payouts in FY 2009 were at a record high of $612 billion, up 2.8 percent from a year earlier and equal to 10 percent of the annual GDP (in 2001 it was 8 percent of the GDP).
5. It is estimated that for FY 2015 payouts will be ¥59 trillion ($733.3 billion) and by FY 2025 payments will total ¥65 trillion ($807.8 billion). This is not sustainable because the productive population is shrinking and the number of premium payments is decreasing.
6. The shrinking workforce and falling tax revenue is aggravating the difficulties that government is experiencing trying to cope with burgeoning costs. It was estimated that for FY 2011 Japan would face a deficit of ¥2,500 billion ($30 billion) in covering payments. In November 2010, Prime Minister Kan was considering tapping the ¥120,000 billion ($1.5 billion) public pension reserves to cover some of the deficit. However, opponents of the proposal stated that the public would lose trust in the system if government paid only half of the basic pension benefits.
7. The public pension premiums fund FILP, which issues bonds (JGB) to IAIs and public works projects and effectively acts as a pay-as-you-go domestic financer of public debt.

According to public opinion polls taken during the Koizumi years, the electorate would have preferred that Koizumi give priority to tackling the social security system. In 2002, as a part of Koizumi's reforms of the public sector, the pension systems of farm, forestry and fishery personnel were consolidated into a single IAI.

In 2004, a major scandal erupted when Koizumi, seven of his cabinet ministers, the president of the DPJ Naoto Kan and 113 Diet members admitted that they had skipped payments to the National Pension scheme. Former Prime Minister Yasuo Fukuda (2007–8), who was the Chief Cabinet Secretary at the time, also confessed that he had withheld payments and resigned from the cabinet. The fact that government leaders had missed premium payments signaled that even they were not confident that they would receive full benefits upon retirement because there would be fewer young people to contribute to the scheme.

A second and much bigger scandal engulfed Prime Minister Shinzo Abe's administration in May 2007 when the DPJ reported that the SIA had lost the records of 50.95 million individuals' pre-1997 payments to the state pension programs when the databases were being integrated. The names, addresses and dates of birth of 50 million people who paid premiums prior to 1997 no longer matched premium payment citizens registered in the public pension system. In addition, 4.3 million records had not been entered into the SIA computer system and 280 local governments in charge of collecting the data had destroyed all of the records. The debacle was attributed to massive bureaucratic incompetence.[14] As SIA staff began a mass exodus from the agency, Abe scrambled to restructure the agency and sectioned it to form six bodies.

Nineteen officials from the MHLW who were employed in the SIA through *amakudari* returned their summer bonuses to take responsibility for the fiasco but six former officials from the ministry and two former chiefs of the SIA refused to do so. The media reported the following October that SIA employees and local government officials had embezzled a total of ¥400 million from 1966 to 2006.

On 11 July, an independent panel was set up to try to assess the credibility of all of the claims but the general consensus was that it would be difficult to identify 38.5 percent of the records and that the former heads of the SIA and former welfare ministers were to blame for the mishap. On 14 March 2008, the government announced that the percentage of the unidentified public pension accounts that had yet to be matched had risen to 20 million, or 40 percent.

Abe's admission in May that he had known about the problems six months earlier served to further infuriate the electorate who blamed Abe and the LDP (whose members were considered to engage in corrupt practices) for the SIA scandal. The LDP suffered a major defeat in the Upper House elections on 29 July 2007, paving the way for Abe's subsequent resignation in September. The DPJ gained thirty-two seats in the Upper House on promises to reform government administration.

According to the 2006 DPJ manifesto, the Government Pension Investment Fund diverted pension premiums totaling ¥370 billion ($3.15 billion) to the construction and operation of resorts located throughout Japan. The resorts were built in the 1980s but dismantled in 2005 due to massive debt. In 2000, the corporation had amassed a debt of ¥2.3 trillion. The state purchased the land, built the facilities and paid for the upkeep, while the Pension Resort Association, together with private and public sector companies, were the operators. Many bureaucrats became executives in the resorts' operators following their retirement.

The government tried to persuade local governments to purchase the resorts in order to alleviate taxpayers of the ¥190 ($1.16 billion) investment in construction but to no avail. The government expected to recover only ¥1 trillion ($8.56 billion) through the sale of the resorts, leaving a debt of ¥11.3 trillion.

The DPJ manifesto also claimed that pension premiums had been used to purchase golf and massage equipment for SIA officials. The party's 2007 manifesto pointed to the SIA as an example of how the ministries used IAIs for post-retirement positions, contending that 27,882 central government bureaucrats were reemployed in 4,576 IAIs and other public corporations.

Public funds disbursed to corporations that accepted these *amakudari* bureaucrats came to 6 trillion yen in a six-month period. Moreover, improper payments to the bureaucrats themselves continued to flow. One example is the numerous past director-generals of the SIA who crossed over to various related public service corporations after they retired, in some cases receiving close to ¥300 million ($2.6 million) in salary and retirement benefits.

On 1 January 2010, the Japan Pension Service (JPS) was established by the MHLW to replace the SIA. The JPS website states:

> The JPS is a special corporation to operate Japan's public pension systems. In order to ensure excellence in public pension operations and to preserve the public's trust, the JPS was established, replacing the Social Insurance Agency which was abolished. The JPS is responsible for public pension systems operations (applications, contribution collection, record keeping, pension consultation, awarding and paying benefits) which are delegated and entrusted by the Minister of Health, Labor and Welfare. The MHLW is responsible for pension finance and administration, while the operation/services are delegated and entrusted to the JPS.

The JPS website also states:

> Our most urgent agenda is to solve the problem with pension records. We need your help in keeping your records correct. If you receive any incorrect or uncertain data, please contact us.[15]

Scandal and super-*amakudari* in a government corporation: Japan Post

Even when he was serving as Minister of Finance in Prime Minister Ryutaro Hashimoto's cabinet Koizumi was committed to privatizing the Postal Service System (PSS). The PSS was established by the Meiji government in 1871 as a part of its modernization policies. The system was based on the British postal service system with postal services, postal savings and postal life insurance provided through the post offices. The postal service offers mail and parcel delivery throughout Japan at affordable rates using an extensive post office network. Postal savings accounts are offered to individual depositors at the post offices. There is a cap of ¥10 million ($122,680) per account. The postal savings business is regulated under the Postal Savings Act and not under the Banking Act and, therefore, cannot offer loans to businesses or individuals. The postal savings cannot participate in inter-bank transfers and depositors are required to transfer money from the postal savings by check.

Until 2003 all of the services were operated directly by the government. In 2003, the three services were reorganized as a single public corporation, the Japan Post Public Corporation (JPPC) and managed by the state. The JPPC was the nation's largest employer with 400,000 employees (one-third of all government workers) and a nationwide network of 24,700 post offices, which provided all of the services. Until it was reorganized again in 2007 JPPC held 25 percent of household assets. It was the world's largest holder of personal assets with ¥224 trillion ($2.1 trillion) in saving accounts[16] and ¥126 trillion ($ 1.2 trillion) in life insurance premiums. The corporation held ¥140 trillion or 20 percent of Japan's debt through JGBs.

Koizumi contended that a privatized JPPC would serve to curb government spending and the growth of public debt. The proponents of privatization also claimed that it would help to eliminate a large source of corruption and pork-barrel patronage as well as allow a greater efficiency and flexibility in the use of company funds. In 2003, Koizumi proposed dividing the JPPC into four companies by April 2007. The companies would include the bank, the life insurer, the postal delivery service and network operations, all placed under a government-controlled holding company. The savings and insurance companies would be privatized and sold by 2017.

But Koizumi's privatization efforts met with stiff opposition from members of his own party, the LDP, and from postal service employees. Politicians relied on the postal savings' significant resources to fund FILP (i.e. public works projects) and many of their constituencies in rural regions were concerned about reduced postal services. Postal service employees, who were an influential lobbying group and who rallied rural voters, were concerned about job losses if the JPPC were privatized. The rural postmasters traditionally collected substantial funds for LDP campaign coffers from the depositors in the postal accounts.

Koizumi's bill was initially voted down in the Upper House of the National Diet in August 2005. Determined to push the bill through the Diet, the prime minister dissolved the more powerful Lower House on 9 August and called a snap election (the Upper House cannot be dissolved). As the president of the LDP he also kicked out thirteen members of his party who were opposed to the privatization, and successfully persuaded friends and colleagues in the public and private sector who had high profiles in the media to stand for election.

Nationwide elections for the Lower House were held on 11 September and Koizumi won the majority by 233 to 228. One of Koizumi's newcomers was Takafumi Horie, the flamboyant entrepreneur and president of the Internet portal operator Livedoor. Although Horie ran as an Independent in the Hiroshima Sixth District, he received full support from the LDP. Horie lost the election and his fortunes continued to decline when he was arrested on 23 January 2006 for accounting fraud.[17] He was released on bail the following April but after a lengthy trial beginning in September, he was convicted and sentenced to two and a half years in prison on 17 March 2007. However, Horie, free on ¥600 million ($5.1 million) bail managed to defer his imprisonment by appealing to the Japan Supreme Court but his appeal was rejected on 28 April 2011 and he was imprisoned for twenty-three months on 20 June.

Koizumi's privatization bill was passed in October 2005. Two years later in October 2007, the JPPC was divided into four companies and managed by the holding company. JPPC was renamed Japan Post (JP). By 2017, the holding company would dispose of all of its shares in the savings and insurance companies, which these entities would be able to repurchase. Similar to the case for JNOC, the government would possess at least one-third of the shares. In other words, the world's largest bank would be partially owned by the Japanese government.

Among the issues raised regarding the future of the JP was whether the Japanese mega-banks would be able to compete in domestic and global markets since the bank would receive support from the state. In expectation of the privatization, the postal bank began selling investment trusts, offering nine funds that were managed by fund houses. It predicted sales would eventually take over a 50 percent share of the financial services market.

Orix Chairman Yoshihiko Miyauchi, the major investor in the Murakami Fund, helped to plan the privatization scheme. The president of Sumitomo Bank Yoshifumi Nishikawa, who also collaborated in the design of the scheme, was appointed the first president of Japan Post Holdings Co. However, the planned sale in late 2008 of a Kanpo no Yado nationwide resort network to Orix Real Estate Corp. for ¥10.9 billion ($106 million) enraged Kunio Hatoyama, who was the Minister of Internal Affairs and Communications for Prime Minister Taro Aso (September 2008–September 2009) and urged Nishikawa to resign at an Upper House committee meeting on 9 June 2009 before Nishikawa's reappointment for another term as president of Japan Post Holdings Co. was slated for approval at a stockholders' meeting at the end of the month. The JP board had agreed in May that Nishikawa would continue for a second term.

The JPPC built the resorts as accommodation for the Postal Life Insurance policy-holders using the life insurance premiums. The construction of the sixty-nine inns and seventy-nine facilities cost ¥240 billion. When JPPC was privatized in 2007, Japan Holdings Co. decided to dispose of ninety-seven hotels that were built to promote health and welfare for the policy-holders because of debts in FY 2003 totaling ¥18 billion and twenty-three facilities established to promote postal savings because of debts totaling ¥10.1 billion during the same period. Japan Post Holdings Co. opened bids for the resort on its website and received bids from twenty-seven companies. According to the JP, two companies, one of them Orix, remained in the running and when the other company pulled out, Orix won the bid with the offer of ¥10.9 billion.

However, Hatoyama was vehemently opposed to the sale because it was at a fraction of the evaluated price of the facilities. In January 2009, Hatoyama pronounced the sale 'a race whose results had already been decided', questioning the competitive nature of the bidding procedure as well as a conflict of interests. Orix later canceled the contract.

Hatoyama, who served as Aso's campaign manager in the LDP presidential race in September 2009, was Aso's key cabinet minister. In June 2009, he threatened to resign if Nishikawa refused to resign. Aso was caught in an untenable position because if Hatoyama resigned, he would become the third minister to depart from his cabinet, weakening the LDP in the coming September elections. If Aso, who had not been in favor of Koizumi's reforms of the JPPC, backed Hatoyama's demand for Nishikawa's resignation he would be perceived as a politician interfering in the operations of a company that was in the process of undergoing privatization.

In the end, Hatoyama submitted his letter of resignation from the cabinet on 17 June. The move was considered by political analysts as Aso's decision to dismiss a colleague who had helped him to become prime minister while consenting to Nishikawa's second term as president of Japan Post Holdings Co., a government-owned company. Hatoyama told reporters on 17 June: 'I don't think that I'm a righteous person; I've made many mistakes. But I just can't tolerate someone who did something dirty and that's my political belief.'

On 30 June, Nishikawa was reappointed by the JP board at a meeting of the general shareholders.

Hatoyama's departure from Aso's cabinet dealt a blow to the Aso administration which had been beleaguered by political scandal and the defection of two other cabinet ministers. On 30 August, the DPJ won 308 of the 480 seats in the Lower House elections, paving the way for Hatoyama's brother, Yukio Hatoyama, to become Japan's first DPJ prime minister. He had pledged a 'drastic revision' of the postal privatization plan, which was opposed by some members of the DPJ.

In March 2010, Kunio left the LDP. There was conjecture that he would join his brother as a member of the DPJ. He also was courted by Aso's former Finance Minister Kaoru Yosano and former Trade Minister Takeo Hiranuma, who had been staunch members of the LDP, to join their new political party, Stand Up Japan.[18] The party platform called for the continuation of the national postal service. Kunio Hatoyama preferred to remain neutral and became an Independent.

The Japanese had high expectations that DPJ reforms would trigger economic stability and growth. However, Prime Minister Hatoyama was in the same predicament as his predecessors. Although the IMF and the OECD urged the Japanese government to raise taxes to help ease the government debt, political gridlock blocked any change in fiscal policies that would wean the economy off public funding. Hatoyama continued the same fiscal policies as the LDP with the release of fiscal stimulus packages and supplementary budgets. Loose monetary policy had become a tradition.

Hatoyama suspended the JP privatization plan, the rationale being that the plan ignored the needs of consumers by focusing too much on profits. Privatization would force the closure of some local post offices, resulting in job losses. Another motivation for preferring to keep the postal service in government hands was the need to finance government spending by returning money from the private sector to the public sector.

It was back to basics and the dependence on pay-as-you-go public financing through the JGB gravy train (i.e. FILP, FLP). At the end of FY 2009 (31 March 2010) Japan Post was reported to have financial assets of ¥304 trillion ($3.490 trillion) or twice that of France's GDP. Japan Post Bank and Japan Post Insurance held ¥303.1 trillion. At the end of June 2010 Japan Post Bank had ¥178.2 trillion ($2.046 trillion) of customer savings, making it the world's largest by deposits. The company invests three-quarters of its assets in JGBs and holds approximately one-third of ¥700 trillion ($8.036 trillion) of Japan's bond market. The government plans to keep one-third of the shares in the holding company – enough to allow a veto of any changes in the company. Essentially, it is a government-run company.

Notwithstanding Prime Minister Hatoyama's promises of implementing administrative reforms, including the *amakudari* system, the appointment of a former MOF official to head the JP Holdings Co. pointed to 'reverse reforms'. Shizuka Kamei, Hatoyama's finance minister, convinced Nishikawa and senior management to resign from Japan Post Holdings Co. by 31 October after the prime minister had frozen the company's asset sales and canceled plans for an IPO of Japan Post. On 21 October, Kamei announced the appointment of Jiro Saito, a former MOF vice-minister (1993–95) to head the postal service. Saito joined the MOF in 1959 after graduating with a degree in law from Tokyo University. He worked in Japan's budget and tax bureaus. At the time of his appointment Saito was the President of the Tokyo Financial Exchange Inc.

Tokyo Financial Exchange Inc. promoted Senior Managing Director Shozo Ota, also a former MOF bureaucrat, to succeed Saito in the post of president. Ota was the fourth former MOF official to preside over the exchange since its establishment in 1989.

Everything old is new again

Special Corporations established prior to 1990 numbered approximately 100 (excluding subsidiaries). In the 1990s there were efforts to eliminate some of the corporations that were debt-ridden through dismantling and through mergers of insolvent with solvent ones. Koizumi succeeded in streamlining the corporations and renamed the remaining fifty-seven corporations Independent Administrative Institutions, indicating that operations would gradually become more transparent. Nevertheless, despite public opinion that the corporations promoted the interests of the ministries and collusion between ministry officials and the private sector, by 2009, the number of IAIs where retired ministry officials found upper management posts had increased to 101. Some of the IAIs have branches throughout Japan and some have both domestic and overseas branches. The freeze on the privatization of Japan Post will ease funding to the ministries' corporations through either FILP Agency Loans or the FILP Loan Fund and *amakudari* in these corporations may become the order of the day. Indeed, posts in the ministries' corporations and affiliated agencies are becoming more sought after as jobs in the private sector decrease.

Although the DPJ manifestos spout the eradication of both the ministries' public corporations and *amakudari*, Hatoyama and Kan found that it was easier said than done, especially during trying economic times. Their administrations were controlled by the same model of administration that controlled the LDP, the party that participated for fifty-three years in the development of a model which is engulfed in the vested interests of politicians, bureaucrats and business owners. The DPJ's win of the majority of seats in the Lower House on 31 August 2009 brought wide media coverage. *The Times* claimed: 'Victory for the Democratic Party of Japan marks an historic shift to a new political era ... Yesterday it [the LDP] faced a modern opposition party with coherent leadership and policies.'[19] *The Economist* called the win 'historic', describing the DPJ as a 'leftist grouping of ruling-party renegades, social democrats and socialists'. Both publications pointed out that Hatoyama's administration might struggle to fulfill the DPJ pledge to reform the bureaucracy while trying to cut public works spending and dealing with rising welfare costs.

Carpenter (2008) stated: 'Even if the Democratic Party of Japan were to be swept into the majority in both Houses by a rising tide of discontent among voters and even if structural reforms are initiated, the effort would be akin to plugging holes in a dyke to stem the flood of problems that are a consequence of years of reliance on an antiquated system and pork-barrel policies.'[20]

3
'Information-Sharing' is Not a Buzz-Word in Japan: Press Clubs Insulate an Insular Political Economy

Foreign journalists covering the nuclear crisis, while expressing admiration for how the Japanese people were coping with the aftermath of the earthquake–tsunami–nuclear crisis, were critical of the way that Japanese government agencies released information about the nuclear accident to reporters. They questioned why there had not been a coordinated effort among agencies to communicate directly with reporters and why the prime minister was not in full control of the flow of data. Although senior officials in government met with foreign reporters during March and Chief Cabinet Secretary Yukio Edano gave interviews to eleven news organizations from the United States, Britain and China, the reporters wondered if the exercise was merely the government's attempt to assuage fears of radioactivity in foods and other products.

The efforts of Japanese and foreign media and commentators to access reliable information is further complicated by the insularity of Japanese organizations and their reticence to open their doors to outside observation. The Japanese social political system is relatively opaque compared to Western industrialized nations and gaining a solid understanding of a given environment can be problematic and therefore time-consuming. This is particularly true of ministerial operations because the ministries are very protective of their territory and tightly control the flow of information. Indeed, ministries are not forthcoming with other ministries and even divisions within ministries are reticent to share data with other divisions.

Foreign observers also face the language barrier. Ministries and government-affiliated agencies may prefer to use special representatives, who are fluent in English, to communicate with foreigners. Some of them have been sent to universities abroad not only to study but also to learn how to interact with foreigners so that they can develop skills enabling them to protect and promote their organizations' interests and fend off criticism. Japanese who do not belong to the institutions may also be accommodated by these representatives. Often interviews are conducted with senior officials who prefer to speak in general terms about their organizations. Furthermore, investigative journalists and analysts have difficulty gathering reliable data concerning ministerial policies because there are no records of court cases, debates or hearings as there are in the United States or the United Kingdom.[1]

Bill Whittaker was the chief correspondent in Japan for CBS Television News from 1989 to 1993. In an interview with the author regarding his experiences in Japan, he called the environment in the United States more 'porous' and spoke about the difficulties in accessing information from Japanese government agencies:

> I do think that the bureaucracy controls the information, not that they manipulate the story but what they choose to reveal and when they choose to reveal it. They play it close to the vest. If they do not want it to get out, it pretty well won't get out. I think that if the bureaucracy decides that the information stays within the bureaucracy, my god, that's where it stays! What I found, also, when it came to the bureaucracy, for the press clubs that operated in different bureaux, if the Secretary [of that particular bureau] said, 'No, you're not going to get that information,' that would be the end of it. From what I understood, if a journalist fell into disfavor within his club and its Secretary, he would get even less information the next go-around.[2]

Press clubs: information cartels control the flow of information

Whittaker was referring to the ministries' press clubs commonly known as *kisha* clubs. The formation of press clubs can be traced to 1890 when Japan's new bureaucracy was established to modernize feudal Japan. Japanese reporters insisted upon entry into the sessions held by the Imperial Diet. Since then, press clubs have served as the major purveyor of information from the ministries and from the Diet to both Japanese and foreign media.

Reporters, commentators and journalists who belong to news organizations and who are covering government activities are assigned to press clubs which are located in offices that are set up to gather news from major organizations such as the ministries, the prime minister's office, political headquarters, local parliament and police headquarters, as well as consumer, entertainment and sports organizations. The organizations offer rooms, telephones and some administrative assistance.

There are now approximately 1,000 press clubs located throughout Japan with the number of members ranging from approximately fifteen to twenty, with the major Japanese dailies regarded as regular club members. The Prime Minister's Club can have over 500 members and the Diet Club can have as many as 5,000 members. Some IAIs such as the Japan Broadcasting Company (NHK), semi-government organizations such as the Nippon Telephone and Telegraph (NTT) and ministries such as the Ministry of Public Management, Home Affairs, Posts and Telecommunications operate their own press clubs. Japan Railroad (JR) and the power utilities have press clubs. Reporters and journalists representing domestic and foreign media are assigned to one club where they may remain during their entire career.

The members of press clubs receive information first. There is a close interpersonal relationship between reporters and their press clubs which prompts reporters to cooperate in the manner of questioning the news source and the way that the information is released to the public. As Whittaker stated, if reporters incur disapproval from the press club they may be sanctioned. They may even be expelled from the club. The Japan Newspaper Association – Nihon Shimbun Kyokai (NSK) – also known as the Japan Newspaper Publishers and Editors Association (JNEPA), plays a fundamental role in choosing which newspapers, journals and television stations can enter a specific club. NSK was founded by Japanese media on 23 July 1946. It is an independent organization based on voluntary membership. The members include the major dailies throughout Japan and television stations, which are also affiliated with the newspapers.

NSK's website states that new members, including news agencies and broadcasters, are chosen by the board of directors and must abide by the Canon of Journalism:

> The prerequisites for membership require that the applying organization pledge to observe the Canon of Journalism as adopted by NSK at the time of its foundation and that it pass an NSK screening to determine whether its recent reporting activities are not contravention of the Canon of Journalism.[3]

According to NSK, as of November 2010 there were 133 members, including 106 daily newspapers, four news agencies and twenty-three broadcasters. The newspapers must have a circulation of over 10,000.

Trade papers and political party organizations cannot enter the NSK. The NSK describes one objective as raising ethical standards and a second objective as protecting media's common interests: 'In the course of NSK history, emphasis shifted gradually to the second objective of protecting common interests, centering on the business activities of newspapers.'

Since NSK decides who enters which press club, accessing information is controlled and, therefore, restricted. The interlocking and vested corporate interests strengthen the press club system because the major newspapers are directly or indirectly linked to broadcasters, namely television and radio stations. Japan has five national newspapers: *Asahi*, *Yomiuri*, *Mainichi*, *Sankei* and *Nikkei*. There are two major news agencies: Jiji Press and Kyodo News. The six broadcast companies are: NHK (an IAI), TV Asahi, TBS (Tokyo Broadcasting System), Fuji TV, NTV (Nippon Television) and TV Tokyo. These organizations tend to control the activities of the press clubs. The direct and indirect institutional connections support the approach of Japanese mass media's conveyance of information to the public and Western journalists often comment on the homogeneous nature of news reporting by Japanese media. Japanese and foreign news providers who are not members of the press clubs experience difficulty accessing primary data.

The NSK's Kisha Club Guidelines state:

> Japan's media industry has a history of applying pressure to public institutions reluctant to disclose information by banding together in the form of the kisha club. The kisha club is an institution and system fostered by Japan's media industry for over a century in pursuit of freedom of speech and freedom of press ... The fundamental purpose of the kisha club system, which has been closely involved with the general public's 'right to know', remains unchanged today.[4]

NSK claims that the press clubs ease access to information from public organizations making fast, in-depth and accurate reporting feasible.

However, the European Commission (EC) in Japan in October 2002 issued *Priority Proposals for Regulatory Reform in Japan*, which called the press clubs 'serious barriers to the free trade of information'. The report stated:

> With the exception of a limited number of wire services (which, if they are members at all, often have only associate membership and, therefore, can listen but have no right to ask questions), membership is denied to journalists from foreign media organizations. It is worth

noting that shukan-shi, or mass-circulation weekly magazines are also excluded, as well as specialized press covering sectors other than those directly related to the host body.

The report delineated the root of the problem of accessing reliable data that is cited here:

1. Officials and the hierarchy of the *kisha* club have the means to prevent the spread of information that they may consider disadvantageous, on pain of exclusion of the offending journalist from the club. The system thus acts against the public interest, since it may deny or delay access to important information, including, for example, information of direct relevance to public health and safety. Reporting of BSE was a case in point.
2. By giving both officials and journalists a vested interest in maintaining exclusivity of a story, the system encourages over-reliance on a single source of information and lack of cross-checking, thus diminishing the quality of information available to the wider public.
3. The system encourages the widespread and undesirable practice of split briefings for domestic and foreign journalists, increasing the potential for information to be tailored to one or the other audience by the briefing party, and exacerbating the risk of spreading inaccurate and biased information about Japan.

The EC threatened to raise the problem with the World Trade Organization (WTO). Although the NSK refuted the statement in its reply to the accusations, Etienne Reuter, a representative of the EC, stated at a December 2003 news conference that 'We reject the statement that our proposals are based on misunderstanding, cultural bias and misconception of facts.'

The NSK acquiesced to the pressure in February 2004 and agreed to allow foreign journalists easier access to government information. Nevertheless, Adam Gamble and Takesato Watanabe contended in their book *A Public Betrayed* that the 2004 reforms were superficial at best. In an interview with Gamble, freelance journalist Tatsuya Iwase condemned the press club system:

> The Japanese Press Clubs are nothing more than transfer devices. They function and will continue to function as mouthpieces for those interests that hold power in the country because all they [press club journalists] do is route transfer of information they are provided by

> news sources. This is assured, because that is the way they are able to remain close to the power they work with.[5]

In his own book, *The Reason Newspapers Aren't Interesting* [*Shimbun ga Omoshiroku nai Ryuu*],[6] Iwase claimed that bureaucrats and government officials found the press club system convenient because the members abided by standards that were set by the government.[7]

The whole truth and nothing but the truth

It is essential for reporters and journalists to belong to a news agency, a newspaper or a broadcaster which is a member of the NSK in order to facilitate entry into a press club. Independent journalists and investigative reporters are at a distinct disadvantage and will struggle to access first-hand information. Accessing a story distributed by other press clubs can be challenging and since there are no records of court cases, debates or hearings, freelance journalists struggle to obtain reliable data regarding bureaucratic policies.

In a Japan Policy Research Institute Working Paper Laurie Freeman claimed that it is common for newspapers to pay members belonging to clubs where they are not given entry to leak information or to write stories for publication. Freeman states:

> The journalists gain a source of extra income and an outlet for some of the information they obtained but cannot write due to club embargoes and other tacit agreements with members of the club or sources.

Freeman concluded that, although this arrangement benefits both the procurer and the supplier, the public suffers the consequences:

> Under this arrangement, because the magazine has gotten such information second or third hand, it either gets written as a rumor or because it is written anonymously and/or does not include an attribution of sources (a practice followed by club members and non-members alike), its news value and believability are seriously diminished.[8]

The DPJ tried to stop the exclusivity of *kisha* clubs by opening the news conferences given by Prime Minister Hatoyama and other cabinet ministers to foreign media organizations and web-based news sites. However, the prime minister's club was still restricted to its *kisha* club members three months after the prime minister had pledged to fulfill his promise. Freelance journalists and non-mainstream news organizations were dissatisfied because there were still far too many barriers. Tetsuo Jimbo, chief editor of the website-based news broadcaster Videonews.com and a former reporter for Associated Press, told *Japan Today*, an Internet news provider, that mainstream news organizations were concerned that Internet news outlets would cut into their advertising revenue which was falling because of the growth of the Internet news businesses.[9]

A meeting was held by academics and journalists on 19 April 2010 to voice an appeal to government and major news organizations to open news conferences to a wider audience. One of the coordinators, journalist Masayuki Takada told *The Japan Times*, another web-based Internet news outlet which collects much of its news from Kyodo News Service: 'We demand unrestricted access to news conferences and conference rooms.'

Calling the *kisha* club system 'a cartel of news reporting', the participants also demanded entry to off-the-record briefings with senior officials because inside information is often passed to major media groups who are determined to maintain exclusive access to information.[10]

TEPCO enjoys a close relationship with major media organizations because of the company's substantial advertising budget, some of which is spent on promoting nuclear energy as a clean and efficient source of power. The media companies are influential members of *kisha* clubs and critics claim that during news conferences with TEPCO's top management regarding the conditions at Daiichi the reporters' questions glossed over vital information such as leaks from the reactors. There were concerns among experts about the possibility of a plutonium leak from No. 3 reactor where MOX fuel was stored because of the hydrogen explosion on 14 March. In an interview with *The Japan Times* on 3 May, freelance journalist and former reporter for the *New York Times'* Tokyo bureau Takashi Uesugi said that he was amazed that reporters did not ask TEPCO about plutonium leaks until 27 March: 'For two weeks, not one reporter asked about plutonium in the press conference. When I raised the question, TEPCO said that it didn't have a detector to check it.'

The following day, TEPCO announced that it had discovered a small amount of plutonium in soil samples taken from the ground at the plant. Uesugi wanted to abolish the *kisha* club system and was one of twenty-two members who established in January the non-profit Free Press Association which pressured the *kisha* clubs to allow non-members to attend news conferences. Uesugi complained: 'Press club members don't want to damage the cosy relationship with TEPCO. This kind of mind-set makes them become soft on TEPCO unwittingly.'[11]

Ironically, *The Japan Times* reported on 6 May that Hajime Iishi, deputy head of the DPJ earthquake and tsunami taskforce, together with two other members of the DPJ took the opportunity to play golf while in the Philippines on official business. Iishi, who was also the vice-president of the DPJ, had been appointed head of the Japan–Philippine friendship parliamentary league in the fall of 2009. Iishi told reporters: 'I thought it would be invisible if it was abroad ... I hope you [reporters] would not do anything that would create uncomfortable feelings among earthquake victims.'[12]

Press clubs and Japan Inc.

Since its inception in 1890 the *kisha* club system has developed into an institution that is controlled through the vested interests of major media organizations, the private corporations, politicians and the bureaucracy. The system is comparable to a mini Japan Inc., paralleling the insularity and rigidity of Japan's system of government administration that was also established in the 1890s. There is an expectation among the foreign media that Japan is an open political society and that accessing political and economic information can be facilitated quickly because the country is the world's third largest economy[13] and appears to have westernized since the end of the Second World War through Western-style democratization. For example, Japanese multinationals have huge outward investment and automobile and electronic manufacturers have achieved astounding success in global markets. But after three years as a correspondent in Japan, Bill Whittaker came to the conclusion:

> I think that what fools you, what is seductive but on the surface is that it looks familiar. You've got tall buildings, freeways, subways and Western dress and all sorts of things that can lull you into feeling that you can relate to it and understand ... But if you're there for a while, you begin to realize that it is just a thin veneer, that the real Japan is behind or beneath that thought, that Western thought.[14]

4
Elements Intrinsic to Japan's Political Economy: Interlocking Interests between an Elite Bureaucracy and Big Business

Prime Minister Naoto Kan was Japan's fifth prime minister within a period of five years. Directly prior to the earthquake, his administration's public approval rating had slipped to 17 percent, putting him in exactly the same position as his predecessors whose terms in office were one year or less. Their administrations were also beleaguered with complex economic and social issues which, since the 1990s, had steadily become more profound.

The DPJ were voted into office because the party had promised extensive reforms of the system. Nonetheless, both Hosokawa and Kan found that the elements in the system that they had promised to reform were firmly entrenched in an antiquated and insular model. This model of administration, until recently, was widely considered to be the reason for Japan's penetration of global markets. However, since government policies had failed to lead Japan's economy out of deflation, promote stable economic growth and shrink the sovereign debt, the model had come to be perceived as an inefficient system that was controlled by the vested interests of political parties, business and the bureaucracy, interests that served to paralyze the decision-making process and to delay the implementation of effective policies.

Although scholars have asserted that METI and MOF were 'the chief architects' of Japan's expanding economic presence in global markets,[1] it can be questioned whether ministerial policies alone were responsible for the resuscitation of Japan's industrial complex after the Second World War and Japan's meteoric rise to become the second biggest economy in the world by 1980. Nevertheless, there is no doubt that the ministries have been the most influential arm of government.[2]

On the other hand, the government's official explanation of the structure of government portrays the prime minister's office as also possessing substantial powers:

> Executive power is vested in the Cabinet, which consists of the Prime Minister and not more than 17 Ministers of State (including Ministers without portfolio and the Chief Cabinet Secretary) and is collectively responsible to the Diet ... [The] Prime Minister appoints the Ministers of State and may dismiss them as he chooses. The Prime Minister, representing the Cabinet, submits bills to the Diet, reports to the Diet on general national affairs and foreign relations, and exercises control and supervision over various administrative branches.[3]

The explanation suggests that the prime minister and the cabinet are at the helm of government. However, Mikuni and Murphy (2002) contend that there is little executive control. They credit Michihisa Hirose, an editor at the *Asahi Shimbun*, for describing the governing system as being controlled by the bureaucracy. Not only are the prime minister and cabinet ministers subject to ministerial guidance but also politicians belonging to the LDP (and the DPJ), who depend on government subsidies for their constituents. Hirose claimed that the subsidies totaled 30 percent of the national budget.[4]

The assertions are realistic. During the past two decades, although there has been an ongoing struggle by politicians to wrest power from the bureaucracy and take control of the policy-making process, there has been no noticeable change in the balance of power. At the height of the political reform movement in 1994 the editorial staff of the *Nikkei Keizai Shimbun* published *The Bureaucracy: A Creaking Giant Power* [*Kanryo Kyodai Kenryoku*]. The book is a compilation of a series of articles that appeared daily on the front page of the newspaper earlier that year and which examined the relationship between the bureaucracy and the Diet as well as the power of the bureaucracy itself. The book explained why bureaucrats were more influential than politicians in Japan's governing system, and why the Japanese did not believe that politicians could plan effective legislation. It attributed their lack of confidence to the following factors: (i) politicians were subservient to the whims of special interest groups; (ii) there was ongoing friction between political factions; and (iii) politicians did not have the expertise or experience to plan effective policies because bureaucrats had been given the power to draft laws since the Meiji period.

The book includes interviews with politicians and bureaucrats that illustrate the ministries' struggle to maintain power during that time. Among the bureaucrats interviewed was Masahisa Naito, the director-general of MITI's Industrial Policy Bureau. When he asserted that bureaucrats operated independently of politicians he was asked if he thought that bureaucrats could work together with politicians to forge policies. Naito felt that it might prove feasible if politicians could plan strategies and the bureaucrats did the legwork, implying that administrators knew more about managing industry and economics than politicians did. He expressed disappointment that, although bureaucrats were the servants of their country, they had become isolated from society and had forgotten their mission.

Naito claimed that the multitude of rules and regulations did not act to support MITI's control of industrial policy but, rather, the trust between bureaucrats and private businesses facilitated the implementation of policy. Naito did not mention the ways that Special Corporations and industrial associations served MITI and the other ministries to persuade businesses to accommodate the policies.

MITI Administrative Vice-Minister Hideaki Kumano was also interviewed. He took exception to the interviewer calling MITI the 'Number Two Ministry of Finance' (*dai-niji Okurasho*), which implied that MITI was taking over the MOF's territory by executing duties that usually fell within the MOF's jurisdiction.[5] The interviewer's insinuation was similar to the allegations by Diet members in 1995 that MITI was the 'Number Two Ministry of Foreign Affairs' as was reported in the 13 June 1995 issue of *Sankei Shimbun*.[6]

The book also gives the views of young officials in the MITI and MOF, who were in favor of the ministries becoming more egalitarian and the relationship between the bureaucracy and politicians becoming more transparent.

Power struggle between politicians and bureaucrats

A classic example of efforts by politicians to poke holes in the bureaucracy and inflame public opinion against elite ministerial guidance occurred in December 1993 when Hiroshi Kumagai, the Minister of Trade and Industry in Prime Minister Morihiro Hosokawa's cabinet demanded the resignation of Masahisa Naito, who was rumored to be in line for the post of International Administrative Vice-Minister.

Kumagai himself had been a MITI official until he resigned in 1976 as director-general of the Small and Medium Enterprise Agency to run for political office on the LDP ticket. He won a seat in the Upper House of the National Diet in 1977 and then took a seat in the Lower House in 1983. In 1991 he served for one year as the Parliamentary Vice-Minister of the Economic Planning Agency before assuming several high-ranking positions in the LDP. He was one of the founders of the New Japan Party with Morihiro Hosokawa in 1992 and served as Minister of State in Prime Minister Hosokawa's cabinet in 1994. Kumagai was also one of the founders of the DPJ in 1998. After a power struggle with Naoto Kan he, along with three of his colleagues, defected in December 2002 to form the tiny New Conservative Party (NCP) where he served as president. After the November 2003 elections only four members remained in the right-wing reformist party. On 10 November 2003 the party accepted Prime Minister Koizumi's invitation to merge with the LDP since Koizumi was also a right-wing reformist.

While Kumagai was in MITI, his relationship with Naito was said to be turbulent. Ostensibly, Kumagai's reason for demanding Naito's resignation was that he had arranged a promotion in MITI for the son of his close friend and patron, MITI International Vice-Minister Yuji Tanahashi.[7] Yasufumi had entered MITI in 1987.[8] The promotion was intended to improve his image and thus his chances of winning a seat in the Diet when he entered the election as an LDP candidate of Gifu Prefecture, a seat that his grandfather, a former governor of Gifu, had occupied for many years. Yasufumi lost the race despite his newly polished image. Kumagai accused Naito, a civil servant, of failing to abide by the principle of non-partisan politics.

The confrontation between Kumagai and Naito was widely covered by the Japanese media. It was reported that Kumagai wanted to stop 'favoritism' in the ministries, using Naito as an example. Naito's dismissal was also related to the power struggle between factions within MITI and between special interest groups in the LDP and the New Japan Party.

When Naito visited Kumagai's office on 16 December 1993, Kumagai told him that he cast darkness over the ministry. When Naito visited Kumagai's office again three days later, Kumagai told him that he could resign 'this month or next month, the process had begun'.[9] Hideaki Kumano, the Administrative Vice-Minister and the head of MITI, was responsible for implementing the dismissal. Although he was against Naito's resignation, in order to keep the peace he pleaded with Naito to resign quietly. Initially Naito refused to resign, calling the stand-off a test of the independence of the bureaucracy. However, on 23 December he resigned without apology, telling reporters that he had simply followed the long-established custom of giving titles to ministry officials who run for office. Although Naito's admission was not welcomed among the general public, who were questioning the integrity of ministry officials, it also illustrated that many ministry officials opted for political careers (as did Kumagai in 1976) and that the ties between bureaucrats and politicians could be very close.

The incident shook the halls of MITI. It was only the second time in the history of the ministry that a high-ranking official was dismissed (the first was in 1952). Naito's subordinates objected to politicians interfering in MITI's affairs and vociferously opposed Naito's resignation. When Naito resigned on 1 April 1994 he moved to Georgetown University as the Marks & Murase[10] Professor in the Asia Law and Policy Studies (ALPS) program. In June, he was reinstated by MITI as a consultant.

Kumano, who had been caught in the middle of the dispute, submitted his resignation from MITI in June to take responsibility for the upheaval in the ministry. Upon his retirement Kumano moved to the Industrial Policy Research Institute (IPRI) where in 1993 Tanahashi had also migrated upon his retirement.

On 7 April 1994, Naito gave a lecture at the Georgetown University Law Center. He reflected on the motivations of ministry officials who worked during Japan's rapid economic growth period and who seemed inspired by their roles as the administrators of Japan's economic rebirth. He lamented the change in attitudes of current bureaucrats, who, he felt, had become inward and 'turf conscious', working to protect their ministry's territory rather than making policy to deregulate markets.

Naito told the audience that the Japanese people, who had relied for centuries on either an emperor or a military regime to govern them, did not want to take the initiative to plan their own destiny but preferred to entrust responsibility to a bureaucracy. He explained that the submission to bureaucratic rule gave the ministries much power which was further enhanced by the close contact between bureaucrats and businessmen, who feared retribution if they did not comply with guidance.

Taking advantage of the increase in the revelations of collusion between ministry officials and big business Japan's major dailies conducted public surveys during the early 1990s to gauge public opinion about the bureaucracy. A good percentage of people questioned by the *Nikkei Shimbun* during 1993 expressed their discontent with elite civil servants, reflecting their reaction to the disclosures of the scandals. The perception of bureaucrats was that: (i) they had a strong elitist mentality; (ii) they were irresponsible;

and (iii) they were clever and shrewd. An overwhelming number believed that bureaucrats should join forces with politicians to plan policies.

The *Nikkei* also surveyed 200 bureaucrats regarding the agencies they considered were no longer necessary or would no longer be relevant in the future. The Hokkaido Development Agency took first place.[11] Among the ministries, MITI took first place. Although the respondents acknowledged that the ministry had done valuable work during the period of rapid growth, they felt that it lacked a clear vision and was groping for an industrial policy that would assist Japanese businesses to internationalize. There was also criticism about MITI's tug-of-war with the former Ministry of Posts and Telecommunications over the administrative jurisdiction of the transportation and high-tech industries and information networks.[12]

The *Mainichi Shimbun* conducted a similar survey in December 1993 with bureaucrats and the general public. Although 77 percent of the subjects questioned credited bureaucrats with Japan's rapid economic growth, 41 percent of the subjects regarded bureaucrats in the 1990s as being greedy for power. Thirty-one percent answered that bureaucrats worked for the benefit of their ministries, and only 18 percent felt that bureaucrats worked for the good of their industrial sectors.[13]

Despite the discontent among the electorate of the most influential arm of government, the system of administration remained contained, defying significant efforts by politicians to initiate structural reforms of the economy.

The origins of the power of the bureaucracy and government links with big business

The Japanese regard their system of bureaucratic administration as having remained fundamentally unchanged since its inception in 1887 during the Meiji Restoration and as steeped in semi-feudalism, despite the end of a military ruler (*Shogun*) whose xenophobic policies effectively isolated Japan from the rest of the world for over 200 years.

This period is known as the Tokugawa era, named for the military leaders who quelled the territorial disputes between the clans to bring peace and stability to the country. Spanish, Portuguese, Dutch and British traders were doing brisk business in Japan when the *Shogun*, fearing that his retainers were being unduly influenced by the interests of some of the traders and that the introduction of Catholicism by Portuguese and Spanish priests would change the values of the Japanese, thus jeopardizing the military's control over the population, closed Japan to foreign trade and prohibited the practice of the Christian religion. Only limited commerce was allowed with the Dutch through Nagasaki, a port in Japan's third largest island of Kyushu.

The government controlled 25 percent of the land and the cities that were considered economically important. The retainer-administrators were known as *samurai*, some of whom also served as soldiers to protect the domains from encroachment by clans from other domains. The government's economic policies focused on controlling and regulating an agrarian economy and increasing production to provide government with as much tax revenue as possible. By 1800, farmers were not allowed to engage in any work other than producing food and if they did not meet their quota, they could be punished for non-compliance (disobedience).

Discontent with the government was already festering when the American Admiral Matthew Perry entered unannounced with his warships into Uraga Bay in 1853 to demand, on behalf of the United States, that Japan open its shores to trade with the United States. The subsequent signing of trade treaties with America, Britain and Holland gave foreigners the right to reside in the port cities specified in the treaties and removed Japan's right to set its own tariffs. The Japanese trade negotiators were the future leaders of the Meiji Restoration government.[14]

From 1854–65, the *Shogun* and his retainers in the western domains tried unsuccessfully to build a defense system against intrusion from the West. During this time, two of the western clans, the Satsuma and Choshu, gained a power base which resulted in a power struggle with the *Shogun* and his other retainers, who were concerned about the growing strength of the two clans. The failure to reach a compromise regarding the establishment of a government that could create an economically and militarily strong nation able to defend itself against occupation by foreign powers triggered a revolt against the *Shogun*, bringing to an end more than 200 years of isolationism.

In 1868, a constitutional monarchy was instituted by members of the Satsuma and Choshu clans and aristocrats. In 1871, a government was consolidated, feudal domains were abolished, prefectures were established under central government control, and the Ministry of Education was established. Toshimichi Okubo, a *samurai* from the Satsuma clan who was instrumental in the overthrow of the military, was one of the key planners of the Meiji government. He and members of the new government traveled extensively around America and Europe from 1871–3 in the first diplomatic mission led by Prince Tomomi Iwakura to collect information about educational systems, technologies and economic and political systems that would assist in Japan's industrialization. The delegation also attempted to renegotiate the trade agreements, which were considered to favor America and Britain.

The consensus was that government-led industrialization would culminate in private initiative as the private sector matured. Sager (2000) states:

> From Okubo's perspective, economic prosperity was intimately tied to national power. The Meiji government's goal was to enrich the Japanese people by helping them as much as possible. The state was committed to the people's prosperity in enterprise, but within the context of serving the national interest as defined by government bureaucrats, not foreign economic theories.[15]

Sager claims that Okubo's policies for Japan's rapid economic development emanated from his 1872 visit to Britain where he studied England's economic development, which had been based on mercantile policies and the protection of domestic markets through laws that gave British ships the right to bring imports into British ports. Okubo concluded that government intervention in the process of industrialization and mercantile policies would spur Japan's economic growth: 'The government would have to work closely with the Japanese private sector to solve real world problems.'[16]

Okubo served as finance minister in 1971, initiating the Land Tax Reform. He was also minister of home affairs and used the ministry to promote industrialization. He abolished some of the entitlements of the *samurai*, such as the right to wear swords in public and the right to receive wages that were generated by rice tax revenues in silver, the currency at the time. Instead, one-third of the wages was paid out in government bonds.

Okubo did not live to realize his industrial policies. When members of the Satsuma clan who were discontented with Okubo's policies rebelled in 1877 against the government's new army (which was established and led by Okubo as the home minister) and were defeated, many of the clan regarded Okubo as a traitor. In May 1878, he was assassinated by members of his own clan.

The Ministry of Public Works was established to aid in industrial development and to produce the same products domestically that were being imported. However, by the late 1870s Japan's trade deficit had spiraled and government spending had escalated. Import substitution policies were terminated and production for export, such as silk yarn and tea, was promoted by the Ministry of Home Affairs in cooperation with local governments.

By the 1880s, Okubo's policies that concentrated on government-led industrial development were being undermined because many of the state-owned enterprises (SOEs) were in deficit. The SOEs were subsequently cut up and sold to large family-owned trading companies who were considered to have the management skills necessary to develop businesses. In return for their cooperation, the companies were granted special privileges by government, thus forging the beginning of an abiding relationship between the government and the private sector.

In 1881, the Ministry of Agriculture and Commerce (MAC) was initiated to institute agrarian reforms, to promote new farming technologies brought from the West, and to guide industrialization. The Bank of Japan (BOJ) was established the following year and was at the foundation of Japan's industrialization. Due to the financing of the new military and the suppression of the rebellion by the Satsuma clan, the government's debt had soared and the resulting inflation prompted the government to consolidate the banking system. It set a limit on the amount of yen notes that the 135 authorized national banks could print. After its establishment the BOJ monopolized the right to issue bank notes. The gold standard was adopted in 1897.

In 1885, the first cabinet system was created. A constitution modeled on Bismarck's Prussian constitution was passed into law in 1889 and a bicameral Parliament was inaugurated in 1890. The first general election was held the following year, culminating in the first Diet (Japan's Parliament).

The bureaucracy's power defined

Japan's government made an all-out effort to achieve economic and industrial equality with the Western powers that had pressured Japan to enter into trade agreements in 1853. The government imported technologies from the West, employing over 3,000 European and American railroad and marine engineers, financial and legal consultants, military instructors, agricultural experts and educators to train the Japanese. Simultaneously, the government sent Japanese abroad to receive instruction in Western methods of production, including those of iron, shipbuilding, and paper, as well as to study the education, military, medical, engineering and legal systems.

Although Japan's economy was undergoing rapid westernization the pre-Meiji military mind-set remained fundamentally unchanged. Aritomo Yamagata is considered to be the major architect of the Meiji bureaucracy. He was a Choshu *samurai* who participated in the overthrow of the *Shogun* in 1867–8 as a staff officer. One of seven political leaders known as *genro*, Yamagata played a number of key roles in Japan's initial military and political development. As the war minister in the first cabinet he laid the foundation of the Imperial Japanese army. And as the home minister (1883–7), he worked tirelessly to suppress political parties as well as labor movements. He was Japan's third prime minister (24 December 1889–May 1891) and also served a second term of office (8 November 1898–19 October 1900).

The persona of the elite bureaucracy was enhanced by Yamagata when, in 1889, he pushed through the Diet the Civil Service Appointment Ordinance regarding the behavior of civil servants that effectively kept political parties from interfering in the bureaucracy. Yamagata's amendment advanced the power of an elite bureaucracy, which he guaranteed until his death in 1922. Yamagata's tenacity led to an autocratic governing system with the bureaucracy dominating the political process. Since that time, policy-making has been guided by elite officials and the legislation written by the elite bureaucracy has been supported by elected officials in the National Diet.

The first Meiji cabinet began the task of educating the new breed of civil servants by opening Tokyo Imperial University in 1887 (now known as Tokyo University or *Todai*) and designing a special civil service entrance exam for the ministries. Since then, graduates from the law faculty in Tokyo University have traditionally had the best opportunity for gaining entry into the elite classes of the ministries and for promotion to the top echelons in the ministries.

Collaboration between the bureaucracy and big business: ministerial guidance and mercantilism (1898–1919)

The construction of infrastructure such as railway lines was vital for the allocation of raw materials and the distribution of goods. Since government funds were constrained and borrowing from foreign banks was limited some of the companies that had purchased the government's SOEs in the 1880s joined again with the government and contributed substantial capital investment to propel Japan's rapid industrialization. This was the case for a number of the large-scale industries, including the development of the heavy industries and shipbuilding during the early twentieth century.

Known as *zaibatsu* ('wealthy cliques'), the large conglomerates' controlling interests were dominated by a single family and engaged in merchant banking, trade, mining and manufacturing. Mitsui's original business in the seventeenth century was based in the kimono industry. In the eighteenth century the company began to engage in money exchange by appointment from the *shogunate*. Sumitomo's business was copper mining in the seventeenth century and the industrialization drive by the Meiji government offered the company the opportunity to import Western technologies and equipment for its mining operations. Mitsubishi expanded its business in 1870 through a government contract to operate steamships.

Cotton yarn was an important early mainstay of industry with heavy investment from wealthy merchants. By 1907, Japan was a major exporter of cotton yarn, as large companies joined to form oligopolies, driving British yarn from the domestic market.

MAC was instrumental in planning industrial development, managing and regulating the production of machinery, shipbuilding and heavy industries. The musket and cannon arsenals built during the Tokugawa era were banned and former arsenal workers were retrained by foreign engineers in machinery production and shipbuilding. The mass production of steel commenced at the Osaka Armory in 1890 with the introduction of Japan's first open-hearth furnace. The government established the steel producer Yawata to accelerate production and the oligopolies followed the government's lead. The Sumitomo Steel, Kawasaki Steel, Kobe Steel and the Nippon Steel Pipe companies were operating by 1912. By 1914, the new steel companies succeeded in producing 44 percent of rolled steel and 64 percent of pig iron for domestic use.

The Industrial Bank of Japan (IBJ) was established in 1900 as a key support mechanism for industrial development to introduce foreign capital. Nevertheless, the government limited foreign direct investment in order to protect immature industries. After the death of Emperor Meiji and during his successor Emperor Taisho's reign (1912–26) great advances were seen in the smoke-stack industries as the collaboration between the ministries and the oligopolies continued to blossom.

Japan's expansion in East Asia, outward investment and the benefits to big business

The Meiji government was determined not to permit occupation by foreign powers. The slogan at the time was 'Prosperous Country, Strong Country, Strong Military'. After several decades of westernization and Western-style education, there was a resurgence of nationalism, Japanese Shinto and Confucius values and the worship of the Emperor.

Japan's initial attempt at expanding beyond its borders was executed in three successive forays. The first was in 1874 in the Ryukyu Islands (Okinawa) which extended from Kyushu to Formosa, a Chinese protectorate. Japan's military invaded Formosa to punish Formosans, who had allegedly murdered Ryukyu citizens when they were shipwrecked on Formosa, forcing China to hand over the islands. A year later, in 1875, Japan seized the Kuril Islands which stretch from Hokkaido to Russian Siberia. In 1876, Japan took the Bonin Islands which are located 1,300 km southeast of the Japanese mainland. The three island chains protected Japan from the north, east and south.

Japan followed the lead of the Western powers' expansion into Asia in 1895.[17] It first pursued Korea, which it wanted to occupy ahead of the other foreign powers. Japan's victory against China in the Sino-Japanese War (1894–5) after a prolonged dispute with China over China's influence in Korean politics and the country's government administration proved the strength of Japan's fledgling military as a force in East Asia. China ceded Taiwan to Japan in 1895 as a part of the Treaty of Shimonoseki and recognized Korea's independence. It also agreed to cede the Liaodong Peninsula in southern Manchuria but France, Germany and Russia pressured Japan to surrender the territory, which Russia occupied soon afterwards. However, Japan's naval presence along the Peninsula helped to establish a foothold in Manchuria.

Japan also was determined to develop Taiwan into a major colony by introducing Japanese education[18] and technologies. The *zaibatsu* invested heavily in infrastructure and manufacturing facilities during Japan's occupation until the end of the Second World War.

A territorial dispute with Russia over Korea's autonomy led to the Russo-Japanese War (1904–5), which Japan won, giving Japan dominance in southern Manchuria. The Treaty of Port Arthur gave Korea to Japan as a protectorate in 1905 and in 1910 Japan annexed the country. The northern region of Korea was rich in minerals and other natural resources, which the *zaibatsu* developed. By 1910, protected by the Japanese army, approximately 170,000 Japanese were living and working in Korea.

Japan also took over the Liaodong Peninsula and the southernmost section of the southern branch of the China Far East Railway. The Japanese military established the Kwantung Army in 1906 to secure the railroad and territory. The partnership between government and the private sector was in full swing.

A semi-private corporation was founded to operate the railroad and develop settlements and industries. The Japanese South Manchuria

Railway or *Mantetsu* employed 35,000 Japanese and 25,000 Chinese by 1910. Through major investment, it became Japan's largest company and by 1916 *Mantetsu* was creating subsidiaries which engaged in steel production, flour and sugar refining, and electric power and chemical plants in Manchuria. Hotels were constructed at each railway station for tourists and warehouses were built for the storage of produce and natural resources to be shipped to Japan. Most of *Mantetsu*'s business was related to the freight of goods for export to Japan and Europe.

Japan's economy and big businesses developed significantly during the First World War. The United States, Great Britain and France used all domestic resources for military supplies, effectively halting the exports of textiles, chemicals and machinery to Japan. Therefore, the Japanese had to increase its production to meet domestic demands. On the other hand, exports to the Allied Powers of military supplies, ships, coal and steel flourished. From 1913–19, total manufacturing output rose approximately 150 percent. Prior to the war, Japan was a debtor nation that imported goods and services from America and Europe. But as the export-led economic boom continued, it became a creditor nation with a trade surplus.

Although Japan saw no action during the war, it took advantage of its position as a member of the Allied Powers to expand into Manchuria. Germany's engagement in Europe reduced the number of troops in Manchuria thus weakening its military presence in the territory which had been leased from China in 1898. In 1914, Japan took over the Marshall Islands from Germany and in the same year invaded the German base of Tsingtao.

On 18 January 1915, Japan submitted to the Chinese government a list of twenty-one grievances, which included the demands that China cease leasing territory to foreign countries, that China grant Japan control of Manchuria and Shantung Province and that it accept Japanese officials as advisors to assist in policy-making. China capitulated to a shortened list of demands on 8 May 1915. However, the treaty was never signed. Although the Treaty of Versailles granted Japan possession of the Marshall Islands, it disallowed the occupation of Shantung, which was returned to China.

The *zaibatsu* along with other large companies entered Manchuria and invested in silver and coal mining, lumber, oil exploration and the construction of more roads and railroad lines to connect ports from where the resources were shipped to Japan for distribution to industry. Idemitsu Oil, one of Japan's major oil companies, is an example of companies that mined fossil fuels and operated refineries in central China.

Recession: the Ministry of Commerce and Industry and industrial rationalization

At the end of the war in 1918 the shipbuilding industry saw a marked decline in orders, a surplus of warships and depressed prices. As orders fell, Japan slipped in the ranks to ninth among shipbuilding nations. The *zaibatsu* countered by diversifying production into automobiles, aircraft and general machinery. The Ishikawa Shipbuilding Company's automobile division became Isuzu Motors Co. and Kawasaki Shipbuilding Co. became Kawasaki Heavy Industries.

Corporate diversification into the heavy industries initiated certain systems that continued during Japan's post-war economic development. The large producers set up their own training facilities to train skilled labor and to ensure that its workers did not migrate to other businesses or join labor unions which had been formed to represent workers and farmers. The seniority system (promotion and wages according to the length of employment in firms) originated during this period.

The post-war *keiretsu* system can also be traced to this time. The smaller firms did not have the capital to finance the training of skilled labor and had to rely predominantly on unskilled labor. Unable to compete with the larger companies, they began to operate as suppliers of manufactured components and goods and services to the conglomerates.

The end of the war in 1918 marked the beginning of a series of recessions and financial crises. By 1919, Japan's economy was booming because of post-war construction activities, expanded exports and new business start-ups. But speculation and an overheated market resulted in a stock-market crash on 15 March 1920 and a panic. The government and the BOJ reacted as the lender of last resort to fund the banks which enabled them to continue to provide loans to key industries.

In the 1920s the *zaibatsu* established holding companies that monitored and audited their affiliated firms. The companies also issued joint shares of stocks. Regardless of the banking crisis, the heavy and chemical industries were developed through joint ventures between Japanese firms and Japanese and American firms. Mitsubishi Corporation was an international player, operating through subsidiaries in Germany and France to import technologies and patents. Mitsubishi Shipbuilding and Mitsubishi Heavy Industries shared about 30 percent patent rights and manufacturing licenses. Mitsubishi Electric signed a contract with Westinghouse to trade technologies.

The Taisho period has been referred to as the 'age of liberalism' because it marked a brief transition of the traditional governing system that was controlled by elite officials towards a parliamentary democracy. Prime Minister Takashi Hara, who assumed office in September 1918, sought to reform the governing system that gave elite officials extra-legal powers to forge and implement policies and cultivate a balance of power between the Executive Office and the bureaucracy. Hara recognized that a fundamental political issue confronting the Japanese was the tension between politicians and bureaucrats (non-elected officials). He attempted to weaken the bureaucracy by dismissing bureaucrats in local governments and other public employees and replacing them with people, regardless of background and rank, whom he regarded as talented and who would contribute to planning long-term economic policies for both national and regional development. The opponents to Hara's reforms were the military and ministry officials from Kagoshima and Yamaguchi Prefectures, regions that were known formerly as Satsuma and Choshu.

Hara allowed peaceful demonstrations by students and journalists in Tokyo calling for universal suffrage regardless of income. The demonstrations were supported by labor unions, socialist, communists and anarchists. Unfortunately, Hara's efforts to promote less government interference in public debate led to his assassination by a railroad worker at Tokyo Station in 1922.

A series of events beginning in 1923 ended the 'age of liberalism'. The economy was dealt yet another blow when the Great Kanto Earthquake hit the Tokyo–Yokohama region in 1923. The earthquake, followed by a tsunami, killed 140,000 people and halted commerce in the region. The government borrowed heavily from the BOJ to implement a ¥400 million disaster relief bill. Due to the imports of massive quantities of goods for reconstruction, Japan's trade surplus disappeared while its trade deficit rose rapidly.

In 1925, to deal with the mounting economic problems, MAC split into two ministries, the Ministry of Agriculture and the Ministry of Commerce and Industry (MCI). MCI authorized recession cartels[19] that covered the majority of industries. It subsidized businesses to promote exports in order to counteract the trade deficit. The recession cartels, the subsidization of businesses and industrial rationalization formed the foundation of MCI's industrial policy, a term used to describe government guidance of economic and industrial development at the macro- and microeconomic levels.

All governments plan industrial policies to support industries that are considered vital to national interests through a set of policy instruments that may include the regulation of production, tax incentives to domestic companies that procure from domestic suppliers, subsidizing of R&D, long-term, low-interest rate loans to businesses for expansion, and protection of industry from foreign competition from imports through tariffs.

The alliance between MCI and big business: the birth of 'administrative guidance'

Emperor Showa assumed the throne upon the death of Emperor Taisho in 1926.[20] The government was able to repay the BOJ by the end of 1926 only 50 percent of the loans that financed the reconstruction, triggering a banking crisis. On 18 April 1927, the Japanese branch of the Bank of Taiwan suspended operations, followed by five more banks on 21 April. The *zaibatsu* took full advantage of the crisis and gained controlling interests in the banks, and took over smaller companies through purchasing equities and placing their executives on the firms' boards. Although the financial crisis directly affected only the banking sector, the economy went in and out of recession which paved the way for MCI to take more control of Japan's industrial development.

Private industry formed the Commerce and Industry Deliberation Council in 1927 to continue until 1930. The organization was the predecessor of the current Industrial Structure Council.[21] With the cooperation of the Council members, MCI planned policy to protect industry from the recession and it was at this time that the concept of industrial rationalization[22] developed. MCI established the Industrial Rationalization Deliberation Council on 19 November 1929, the year of the Great Depression which was sparked by the Wall Street stock-market crash. Nevertheless, by 1929 Japan had become an industrialized power with a GDP per capita equal to the Western powers.

On 2 June 1930, Shinji Yoshino, who was the head of MCI's Industrial Affairs Bureau, set up the Temporary Industrial Rationalization Bureau (TIRB) to operate semi-autonomously of MCI. The TIRB planned policies for the control of businesses, set industrial standards and provided subsidies to support consumer consumption and the production of goods.[23]

MCI expanded its control over industry through the Important Industries Control Law, which was drafted by members of the TIRB and passed by the Diet on 25 February 1931. The law gave MCI the right to form cartels in twenty-six industries considered vital to national interests such as shipbuilding, iron and steel, electric machinery, rayon, paper, cotton spinning, sugar manufacturing and beer brewing.[24] The law laid the foundation for the use of the policy tool 'administrative guidance' which was used consistently by MITI to issue directives to industries to form cartels, to fix prices and to regulate production during Japan's post-war development.

Yoshino became the vice-minister of MCI in 1931. One of his protégés was Nobusuke Kishi, who entered MCI in 1920 after graduating from Tokyo Imperial University with a law degree. Kishi was an assistant section chief in the Secretariat when Yoshino sent him to Germany in 1931 to research German industrial rationalization policies. By 1935, the talented and ambitious Kishi was a top official in the ministry.

The Manchurian Incident

The depression profoundly affected Japan's economy and threw the country into political turmoil, the prelude to militarism and the second Sino-Japanese War. Even before the depression there had been military action in China when Prime Minister Giichi Tanaka (1927–9) sent troops to China to obstruct Chiang Kai-shek's Kuomintang Party's efforts to unify the country. Although the action was unauthorized by government, in June 1928, Japanese troops from Guangdong Province set about protecting Japanese interests in Manchuria.

The Manchurian Incident was the beginning of fifteen years of military and bureaucratic dominance of Japan's government. By 1930, *Mantetsu*'s assets were worth ¥1 billion and Japanese companies with substantial investments were dependent on the army to protect their operations. The Kwantung Army, concerned that Japanese companies would be threatened by the Kuomintang, exploded the railroad tracks outside of Mukden in September 1931. The Kwantung used it as an excuse to place the blame on the Chinese and establish a puppet state in Manchuria and eastern Inner Mongolia in 1932, naming it Manchuko. When Prime Minister Takayuki Inukai (1931–2) tried to cut government expenditure by reining in the expanding military budget he was assassinated by right-wing military officers. A small number of the officers who took part in the assassination were tried and imprisoned but they also were regarded as patriots.

Regardless of protests from politicians, the military continued its forays into Shanghai, engaging with the Chinese, ostensibly to protect Japanese interests in Manchuria. Although the government formally opposed the military action, the army received popular support.

The military aggressively planned the economic expansion of Southern Manchuria and established state-run companies that focused on one industry such as automobiles, airplanes, oil refining and shipping. Nobusuke Kishi served as a key economic planner of industrial development in Manchuko in 1935. He asked MCI to send officials to assist him, specifically TIRB official Etsusaburo Shiina who was three years his junior. Their relationship played an important role in Japan's early post-war political and industrial policies. General Hideki Tojo, who was in charge of military operations in Manchuria, supported Kishi in Manchuko. When Tojo became prime minister in 1941, he appointed Kishi as cabinet minister of commerce and industry.[25]

In March 1935, Koki Hirota, who was foreign minister in Prime Minister Makoto Saito's cabinet, engineered the Soviet Union's sale of the Great Eastern Railway to Japan. Stalin handed over to the Japanese the railroad for $39,200,000 plus $8,400,000 for severance wages to Soviet railway men. The price was considered a pittance because the railroad was worth $400,000,000 in 1905 and over $500,000,000 at the time of purchase.[26] Hirota also established a Japan–China–Manchuko bloc, which was an organization designed to suppress communism and anti-Japanese activities in China.

In 1937 Kishi invited his relative, Yoshisuke Aikawa, the president of Nissan *zaibatsu* to relocate to Manchuko and establish a 50–50 joint venture with the new Manchuko puppet government. At first Aikawa resisted because Manchuria did not have the necessary infrastructure or subcontractors needed to develop the automobile or aircraft industries. Nevertheless, with guaranteed loans from the IBJ and the BOJ, Aikawa established the Manchurian Heavy Industrial Development Corporation. *Mantetsu* and other Japanese industrial firms in Manchuko were connected to the company through cross-share holdings. American steel companies also financed loans. Japanese companies thrived in Manchuria, Korea and Taiwan until the end of the Second World War.

Fascism came to the fore on 26 February 1936 when 1,400 junior army officers attempted a *coup d'état*, seizing key government buildings in Tokyo and assassinating three high-ranking officials. However, they were foiled in their attempt to kill Prime Minister Admiral Keisuke Okada, whom the military resented because of his support of a democratic parliamentary government. Okada resigned and Hirota became prime minister in March 1936. During his administration, the Anti-Comintern Pact with Nazi Germany and fascist Italy was signed. Hirota resigned after eleven months in office over a disagreement between the cabinet's war minister and politician Kunimatsu Hamada.

The Second World War: the intensification of ministerial powers

The military budget was increased substantially in 1936, following which Japan declared war on China in July 1937. The advent of the war gave the economic ministries, MCI and the Ministry of Finance more control over the industrial and banking sectors and a closer alliance with the *zaibatsu* who produced armaments and military supplies. A close partnership was an inevitable result of the all-out effort to win the war in the Pacific. The major *zaibatsu* operating in 1937 were Mitsui, Mitsubishi, Sumitomo, Furukawa, Nissan, Yasuda, Okura and Asano.

Emergency legislation was enacted, which allowed the bureaucracy to mobilize the economy. In 1937 the pro-military cabinet pushed three laws through the Diet that effectively gave the state extensive controls:

1. The Export–Import Commodities Emergency Measure Law gave the munitions industry priority for the allocation of vital raw materials and chemicals.
2. The Emergency Capital Allocation Law was used to give the munitions industry priority treatment, controlling the establishment of firms, dividend payments, capital increases, bond flotation and loans.
3. The Munitions Industrial Mobilization Law extended bureaucratic command over production.

In 1938, the General Mobilization Law was passed, giving the state complete control over political society. The three institutions that played major roles in administering the wartime economy were MCI, the MOF and the BOJ. MCI managed and regulated industry as a whole, as well as munitions production. MCI merged with the Cabinet Planning Board on 1 November 1943 to form the Ministry of Munitions (MM) in order to improve the production of military supplies. Although Prime Minister Tojo was MMs first vice-minister, his deputy Nobusuke Kishi oversaw operations. Kishi had appointed Etsusaburo Shiina as the director of the all-important General Affairs Bureau in 1939 and Shiina appointed the officials who were ambitious and control-orientated to serve under him.

The BOJ poured funds into both of the banks that were established by MCI and the MOF and to private banks (the *zaibatsu* banks were the largest) to finance the production of goods considered vital for military supplies, particularly munitions. The Industrial Bank of Japan[27] administered by MCI and the MOF was a crucial support mechanism, providing funds for munitions. Motor companies such as Toyota, Nissan and Isuzu produced military vehicles.

In 1944, the 'System of Financial Institutions Authorized to Finance Munitions Companies' was established, the bill giving the government the authority to order private banks run by the *zaibatsu* to fund munitions manufacturers, thus ensuring a steady supply of armaments. Each bank was assigned to one munitions firm. In 1945, the MOF established two banks to allocate funds to over 600 firms, assigned to fulfill their production quotas. The banks were called 'Financial Institutions Authorized to Finance Munitions Companies' and this system continued as the 'main bank' system until the 1990s. The banks were protected against risky lending through government loan guarantees. By 1945, regardless of whether firms engaged in the production of military supplies, 2,000 had been assigned a bank that would be in charge of their financing. The post-war formation of the giant financial groups known as *keiretsu* was a consequence of this close relationship between the *zaibatsu* and the 'main bank' system.[28]

Enduring changes: corporate culture

The mobilization of the wartime economy brought about some important changes in the corporate culture that have remained relatively unaltered throughout Japan's post-war era:

1. Trade and industrial associations, which had originated in the 1920s, became popular during the war as MCI's controls over industry intensified.
2. In the 1930s, groups of corporations similar to the *zaibatsu* became prevalent, which increased the powers of corporate management because the new shareholders were eager for high-growth. Shareholders' interests were eliminated, managers were given more power, and employees were motivated through company unions and job security. Trade unions were abolished and substituted with corporate in-house union activities. This effectively reduced the risk of strikes and the disruption of production.
3. Businesses employed bureaucrats to forecast government directives.

Reverse reforms and Japan Inc.

Mikuni and Murphy evaluate Japan's central government as possessing little real power, describing it as a 'weak confederation of ministries' acting as 'sovereign entities', each administering and acting for segments of society. The authors explain that elite bureaucrats are today's policy-makers, who operate independently of legal sanctions, thus giving them unlimited power.[29] The national ministries draft most laws, ordinances, regulations and licenses. They can operate, in general, unconstrained by the judicial system.

The bureaucracy was endowed with the legal capacity to operate free of political interference during the Meiji period. However, it was during the war, through state intervention in the Japanese economy, that the ministries' institutional mind-set of management of Japan's post-war economy was established. The extra-legal authority that today's bureaucrats enjoy is connected to the wartime system of government administration.

The United States' policies in Japan after Japan's defeat in the Second World War were fundamental to the support of what has become known as Japan Inc. and Japan's post-war bureaucracy as the most influential arm of government.

The period 1945–52 is commonly referred to as the Allied Occupation. The occupying countries were the United States, Great Britain and Australia but Great Britain and Australia sent far fewer troops than the United States, which was commanded by Douglas MacArthur, the Supreme Commander of the Allied Powers (SCAP). The objectives of the occupation were three-fold: (i) Japan's demilitarization; (ii) the purging of war criminals; and (iii) Japan's democratization and economic resuscitation. The United States was determined to create a Western-style capitalistic and democratic society.

SCAP intended to reform Japan's wartime economic system by systematically disbanding the military and the ministries. The Ministry of Munitions, along with the Home Ministry and the Military Police were abolished in 1945. Policies also included the dismantling of the *zaibatsu* because of their full participation in the war effort, initiating a new constitution to introduce democratic principles such as freedom of speech, free elections and female suffrage. Labor unions were allowed to reorganize and the Socialist and Communist Parties that had been banned during the war were also given the right to reorganize.

The occupation was scheduled to end in 1948 with a Western-style constitution and democratic governing system in place. A new constitution, which removed the Showa Emperor as the formal head of state, replaced the Meiji constitution in 1947. The famous Article 9 renounced Japan's right to wage war.[30] The Ministry of Defense was abolished. Wartime government officials were purged as Class A war criminals and executed.

Among the executed were Tojo and Hirota. Kishi was purged as a Class A war criminal and imprisoned in Sugamo prison in Tokyo.

However, the beginning of the Cold War in 1945 with the Soviet Union and the perceived threat of Communist expansion in East Asia, the Communist-backed General Strike by the Government Workers' Union in 1947 and the Korean War (1950–3) persuaded SCAP to reconsider a number of its intended reforms in order to cultivate an economically strong and politically conservative ally in the Pacific, where it could base its military operations and hardware. In 1947 the Diet banned strikes by government workers and in 1948 the Wage Control Program replaced free collective bargaining.

SCAP recognized that fundamental to a swift recovery for Japan's devastated industrial complex was the expertise of the ministry officials who had managed Japan's wartime economy. It reinstated former ministry officials, thereby preserving Japan's pre-war institutions and the economic system. SCAP's staff ignorance of the Japanese language and social system compelled them to rely on existing institutions, namely the bureaucracy, to implement policies. The nature of bureaucratic rule and the persona of the ministries remained intact because the officials who planned Japan's post-war industrial recovery were former MCI and MOF officials.

Forty-two officials from the MM took upper-management positions in MCI, which was reestablished in 1946, and then in the Ministry of International Trade and Industry (MITI), which was formed through the consolidation of MCI and the Board of Trade in 1949. Officials who served under Etsusaburo Shiina in the General Affairs Bureau in MCI, including Takayuki Yamamoto and Tomisaburo Hirai, Tenjiro Yoshida, Takeo Ishihara and Hisatsugu Tokunaga, became MITI vice-ministers in the 1950s and 1960s.[31] They planned Japan's initial industrial and economic policies while maintaining a close relationship with their former MCI colleagues.

SCAP dissolved the *zaibatsu*'s holding companies so that the family owners no longer held the controlling interest. The 'economic purge' of the executives who had directed the companies during the war was delayed until January 1947, when ultimately 1,400 executives were retired. The Mitsubishi and Mitsui trading companies were dismantled in July 1947 and Nissan was dissolved, even though Aikawa had vehemently opposed the Tripartite Alliance and, consequently, had been forced by the army to resign his post as *Mantestu*'s chairman and return to Japan in 1942.

Out of 325 companies slated for liquidation only eighteen were actually split up because the Cold War convinced SCAP to reverse its initial stand and follow a more lenient course, sanctioning the return of the conglomerates to engage along with the ministries, mainly MITI and MOF, to rebuild the economy.

The reforms of the *zaibatsu* were negligible because the conglomerates reorganized into groups of enterprises (i.e. *keiretsu*) with the *zaibatsu* main banks at the center of the capitalization. The financial groups that funded the domestic economy included Mitsui, Mitsubishi and Sumitomo. Centered on one bank, a 'main bank' that funded corporate investment, the corporations had cross-shareholdings and directors in common. Smaller firms were on extended contracts to supply parts and services. Some of the firms were partially owned and produced parts solely for one *keiretsu*. They were known as 'first-tier suppliers'. The 'second-tier' and 'third-tier' firms were contracted to supply smaller amounts and produced for other companies as well. Many of the firms had been suppliers to the *zaibatsu* during the war.

The Anti-Monopoly Law was enacted in 1947, but relaxed by 1953 to allow the financial companies within each conglomerate to own as much as 10 percent of the shares of the firms not connected to finance. The law also allowed MITI to form import–export cartels, price cartels and production cartels, legally sanctioned by the Japan Fair Trade Commission (JFTC). In order to increase production capacity, MITI's Industrial Rationalization Council requested that the *keiretsu* invest capital in domestic firms.

The BOJ initially controlled net additional lending of metropolitan banks and by 1957 of regional banks, dictating to each bank quarterly the amount acceptable through the extra-legal policy tool 'window guidance' (*madoguchi shido*). The BOJ decided the quota according to the ranking of the banks from the metropolitan banks down to the regional banks. The state banks were also ranked. The IBJ ranked first and assumed an even more important role that it had played during the war, providing low-interest rate loans to firms to support rapid industrial growth. The Long-Term Credit Bank and the Nippon Credit Bank (see p. 34) were ranked second and third respectively. Banks adhered to the dictates because through this policy, they could achieve an equal market share and simultaneously be protected by the BOJ, as the lender of last resort, and MOF, the ministry that protected the banks against losses. 'Window guidance' also served to pressure banks to hire retired officials. Werner (2003) refers to an incident involving a bank in Nagoya which dismissed a BOJ *amakudari* official. The dismissal was considered a form of 'misbehavior'. The BOJ would 'punish' banks by reducing the loan quotas.[32]

After the war and prior to 1955 the Liberal Party was the main political party. Its president Shigeru Yoshida served as Japan's forty-fifth prime minister (May 1946–May 1947) and Japan's forty-eighth prime minister (October 1948–December 1954). A former official in the Ministry of Foreign Affairs, he was Japan's ambassador to Italy and to Great Britain during the 1930s. He was imprisoned briefly in 1945. Yoshida's policies emphasized the rebuilding of Japan's industrial complex through rapid economic growth while depending on the United States military alliance which allowed Japan to pour all its resources into the resuscitation of the economy. Known as the Yoshida Doctrine, the policies impacted significantly on Japan's future domestic and foreign political economic policies.

The San Francisco Peace Treaty, signed on 8 September 1951 between Japan and forty-nine countries, officially ended the war, divesting Japan of occupied territories and confiscating Japan's foreign assets. The US–Japan Security Treaty, signed a month later, ensured an economic and military alliance between the United States and Japan.

Japan's economy benefited considerably from the Korean War because the production of munitions for the American military increased exports by 26 percent. Between 1955 and 1973, Japan's GDP increased almost six times, or 11.5 percent. The MOF, MITI and the Economic Planning Agency's policies were considered to be the drivers of rapid economic growth. Nevertheless, it is difficult to quantify the extent to which ministerial policy actually drove the economy. Besides SCAP's 'reverse reforms' other factors should also be considered:

1. A cheap currency until 1971. Known as the Dodge Line, the yen was set at the cheap rate of 360 yen per dollar to assist Japanese exports.[33]
2. America's open markets.
3. Low military spending.
4. Protection of domestic industries from foreign competition with tariffs, quotas, investment controls and bureaucratic red tape.
5. Access to cheap technologies, especially from the US.
6. Prime Minister Hayato Ikeda (1960–4) whose 'double income' economic policies inspired the Japanese. Focusing on the development of the heavy industries as opposed to small businesses, he predicted a 7.2 percent rate of growth within ten years. The rate of growth was 11.6 percent by the end of 1960.
7. The 1964 Tokyo Olympics marked a spurt of infrastructure construction and industrial development.

8. Consistent support for policies from both the National Diet and business.
9. High savings rate.
10. Industriousness of a literate, skilled workforce.

The ministry officials refused to capitulate spiritually to the American Occupation and worked vigorously to plan and implement policies that would revive Japan's economy. Japan rising from the ashes of defeat and overtaking the United States on the economic front would demonstrate to the world that Japan was a power to be reckoned with. Werner (2003) states:

> for all intents and purpose Japan's wartime economic control remained unchanged after World War II. The very bureaucrats and managers, who had demonstrated excellence in running the fully mobilized economy, whether in Manchuria or back home, received rapid promotion in the post-war system.[34]

By the time Japan had achieved the third highest GDP in the world in 1970 (behind the United States and the Soviet Union respectively), the old-style bureaucratic system was firmly entrenched. The officials' loyalty to their ministries and the ministries' reticence to release control of their administrative territories and deregulate markets instilled insularity and rigidity in the administrative system.

Small shock, big shock: nuclear energy as a national priority

The end of the Bretton Woods system of fixed exchange rates and the gold standard on which currency standards were based was referred to by the Japanese as 'the Nixon Shock'. In the early 1970s, the Vietnam War caused a rapid rise in inflation in the United States, which was also experiencing a trade deficit for the first time in the twentieth century. The dollar that had been valued at $35 per ounce of gold was weakened by America's inability to cut spending and reduce its trade deficit. As the government printed more money to fund operations in Vietnam, America's gold coverage of the dollar dropped to 11 percent and in August 1971, President Richard Nixon imposed ninety-day wage and price controls as well as a 10 percent import surcharge. This action effectively made the dollar inconvertible on the open market. Even though the import surcharge was dropped five months later, there was a general reevaluation of currencies. By 1976, there was no longer a fixed exchange rate.

The MOF suppressed the yen, which had been ¥360 per dollar for twenty-two years, to ¥300 per dollar so that prices of exports to the American market were relatively unaffected. MOF officials felt that as long as Japan ran a trade surplus with the United States, they were in control of the economy and their administrative territory. Japan continued to hoard dollars in the United States and to control the yen in Japan.

MITI's long-term industrial policies concentrated on the development of the heavy industries, namely the petrochemical industry and other energy-intensive industries such as aluminum and plastics. However, these policies were virtually ended by the 'oil shock' which began in October 1973 when the Organization of Arab Petroleum Exporting Countries (OAPEC), consisting of OPEC, Egypt, Syria and Tunisia, announced an oil embargo to protest the United States' military support of Israel during the Yom Kippur War. The embargo lasted until March 1974 but it was considered by many countries, including Japan, as instigated by the United States. In January 1974, when Prime Minister Kakuei Tanaka (1972–4) met with Henry Kissinger he was unable to receive a guarantee from the United States for a steady supply of crude oil in return for Japan's support of American policies, thereby pressuring Japan to take a pro-Arab stance. Japan was recognized by the Arab nations as pro-Arab. Although the embargo ended in March the price of crude oil rose from $3 to $12 per barrel and the yen appreciated to ¥260 per dollar.

Steel, automobiles, heavy and home appliances, petrochemicals, autotires, synthetic fibers, aluminum, non-ferrous metals, plate glass, pulp and paper were industries that were among those directly affected by the oil embargo and increase of petroleum prices. The oil crisis forced the Japanese government to recognize the country's vulnerability and it took immediate measures to rationalize oil in order to decrease the demand for both gasoline and electricity. Tanaka's cabinet, which included Minister of International Trade and Industry Yasujiro Nakasone, implemented the policy instrument 'administrative guidance' for 'the Conservation of Oil and Electricity of Oil and Electricity in the Private Sector'.

MITI had been using the tool formally since 1952 in order to form cartels of industries designated as important to national interests, protecting them from foreign competition. The tool is used *ad hoc* and at the discretion of the ministries. There are no laws that limit the number of times the guidance can be used, giving the ministries uncommon powers to regulate. Companies will usually receive notification requesting that they follow ministerial regulations. The directives are either transmitted in writing or by telephone, although a law in 1997 officially curtailed the use of the telephone.

MITI's ANRE drafted the Petroleum Supply and Demand Normalization Law (PSDN) which was passed by the Diet in December 1973. After the embargo was lifted in February, 'administrative guidance' was replaced by PSDN. The law gave MITI control over the supply and consumption of oil to both large and small companies and oil producers and dealers reported their production and sales to MITI. MITI could direct companies to change plans and to keep supplies in stock or to sell. MITI could 'guide' the Association of Petroleum Dealers to secure the national oil supply (see p. 42). Production and price cartels were included in MITI's control policies. In 1973, the government announced that nuclear energy was a national priority.

MITI began to restructure industry, focusing on high-tech information systems technology and telecommunication sectors. By 1978, the structurally damaged energy-intensive industries such as petrochemicals and synthetic fiber were considered 'sunset' or 'slump' industries.

Mechanisms to implement guidance

MITI continued to maintain substantial control of its administrative territory through protectionist policies. The policies included doling out subsidies to the depressed energy-intensive industries and the continuation of the policy instrument 'administrative guidance' to form anti-recession, rationalization and price cartels under the Structurally Depressed Industry Law. The cartels had been used during the 1920s and during Japan's post-war rapid growth period in order to prevent destructive competition between producers (i.e. keep a level playing field) and to protect domestic markets. The cartels played a primary role in limiting competition from foreign manufacturers. Yamamura (1982) maintained that the anti-recession cartel policy was merely a continuation of the policies that had promoted rapid economic growth and that the management of capital investment, while reducing risks through the use of cartels, was counter-productive. The LDP supported MITI's policies because politicians needed contributions from businesses for their campaigns. Yamamura also contended that the continuation of these policies benefited MITI officials as well:

> The LDP was understandably anxious to accommodate the wishes of the industry since the latter provided literally all of the LDP's political

> funds ... some high-ranking MITI officers had an incentive to be more helpful to the largest firms that had invited them to fill high managerial posts after their retirement.[35]

'Administrative guidance' was easily facilitated through *amakudari* which served to connect retired ministry officials employed in the companies that were receiving directives and MITI.

Katz (1998) claimed that cartel-like policies in the 1990s were continuing to protect domestic markets even though the government insisted that many of the import barriers used in the past, including tariffs, had been eliminated. He believed that members of industrial associations, such as petroleum, colluded to fix prices while the JFTC turned a blind eye. Informal guidance may be given by the ministry through trade and industrial associations, whose members feel obliged to follow policy, such as purchasing solely from domestic producers. Katz also claimed that companies feared retaliatory measures if they did not comply, referring to Mitsubishi Heavy Industries as an example of a company that dared to purchase cheaper steel from a Korean firm at a third of the cost of Japanese cartel-fixed prices. Mitsubishi experienced a cut-off of some supplies and other penalties.[36]

Among METI's oil and gas-related industrial associations are:

1. Petroleum Association of Japan (PAJ)
2. Japan Cooperation Center, Petroleum (JCCP)
3. Japan Petroleum Institute (JPI)
4. Japan Petroleum Energy Center (JPC)
5. The Japan Gas Association

Among METI's electricity-related industrial associations are:

1. The Federation of Electric Power Companies of Japan (FEPC)
2. Electric Power Development Company (J-POWER)
3. Japan Heat Service Utilities Association (JHSUA)

Mikuni (1998) explained the role industrial associations played in the formation of cartels, as they had done in the 1930s:

> These industrial associations and the implied promise of government intervention had helped maintain the viability of member companies, thus obviating the need for individual companies to accumulate

> the sufficient capital on their own to withstand downturns. Anti-trust laws do exist, but they are largely empty.[37]

It is interesting to note that a number of JFTC board members have been MOF officials. In 1997 the holding companies that SCAP disbanded in 1947 were reestablished. Cross-shareholdings in the *keiretsu* were again sanctioned.

Mercantile policies continued throughout the 1980s and 1990s. Mikuni contended that Japan's economic problems were caused primarily by these policies. He and Murphy warned in 2003 when Japan was the largest creditor in the world with the largest trade surplus:

> A mercantile regime may enjoy a period of success as it accumulates specie – or, in today's terms, claims on other countries. But it usually runs into some kind of monetary train wreck. The mercantilist's efforts to convert its claims on others into its own currency drive up the exchange rate because foreigners, whose sales to the mercantilist country are restricted, have few ways of accumulating the mercantilist's currency in the first place, so little of its currency is held outside its borders. The mercantilist exporters then find that no one can afford to buy their goods.[38]

5
The DNA of Japan's Post-war Political System: Ultra-conservative to the Core

The Japanese described their post-war governing system as a 'ruling triad' of elite bureaucrats, conservative politicians and leading businessmen. The Liberal Democratic Party dominated Japanese politics from 1955 to 2009 with the exception of a brief three-year break from 1993–6. In most Western countries, no single party has been in power long enough to give bureaucrats the consistent support to draft laws and implement policies nor are there democratic societies where ministries can operate unfettered by legal sanction as they operate in Japan. There are three key reasons for this unwavering support:

1. Japan's political economic system can be characterized as pork-barrel and protectionist. Big business and business federations made large contributions to LDP coffers. The LDP received votes and large donations from traditional support groups, such as small local firms and from businesses engaged in construction, transportation and telecommunications, in exchange for public works projects.
2. The LDP received substantial support from special interest groups represented by the ministries *vis-à-vis* industrial associations and federations.
3. The network of bureaucrats throughout Japan's socio-political system. Bureaucrats traditionally have sought political office in both national and local government Diets and as governors and vice-governors in the prefectures. When the snap election was held on 11 September after Koizumi dissolved the Lower House in 2005 ninety-five former government officials ran for seats, fifty-seven on the LDP ticket and twenty-five on the DPJ ticket. Koizumi himself recruited an elite career official from the MOF to stand in the election. Satsuki Katayama had entered the MOF in 1982 and achieved a number of influential

positions in the ministry, including Director of the Policy Evaluation Office, Director of Overall Coordination Division in the ministry's Secretariat in 2000, and Director of Legal Affairs in the Budget Bureau in 2004.

Due to SCAP's purges of Japan's wartime politicians, the conservative parties that existed prior to the end of the war lost most of their members and there was a scramble among the remaining party members for power at the end of the war. The mainstream Liberal Party leader Ichiro Hatoyama was slated to become prime minister in 1946 but was purged by SCAP and prohibited to enter politics for five years until 1951. During his absence, Shigeru Yoshida became the president and assumed the office of prime minister for two terms (1946–7 and 1948–54). When Hatoyama re-entered politics and the Liberal Party, Yoshida refused to stand down as the party's leader.

Former MCI official Nobusuke Kishi (1896–1987) is credited with being the father of the LDP and, also, the kingpin of the 1955 political system that was based on the collaboration between politicians, businessmen and bureaucrats and banked with huge sums from the private sector. His career is indicative of how SCAP's 'reverse reforms' effectively perpetuated Japan's wartime government administration of the economy and an ultra-conservative political environment which further enhanced the power of the bureaucracy. One of Kishi's heroes while he attended the university was Kita Ikki, a right-wing radical who promoted military rule and the conquest of Manchuria and Siberia.

Kishi was a power-broker in Manchuria where he skillfully maneuvered the distribution of capital to cultivate right-wing nationalists and allies in non-*zaibatsu* and small/medium-size businesses. Kishi was an expert at laundering money no matter how large the amount. It was rumored that Kishi, through his connections with the opium trade, engaged in legal as well as illegal transactions for both public and personal purposes through a simple telephone call.[1]

Before the outbreak of the war, Kishi formed the Association for the Defense of the Fatherland. When he returned to Tokyo in 1939, he extended his network and established in 1944 the 'Kishi New Party' (*Kishi Shinto*). The members were Kishi's political and business right-wing associates from Manchuria and elite control bureaucrats and right-wing nationalist military personnel who had planned the 1931 *coup d'état*. His money power base were the directors of the public companies in Manchuria and the non-*zaibatsu* independent businesses that had reaped profits through Kishi's agendas.

While in prison, Kishi reflected on his past and the war. Although he had served the Emperor faithfully and Japan's war machine without reservation, he concluded that the war had been futile and that Japan should never be involved in a war again. When he realized that he was not going to be executed he was determined that the rest of his life would be spent effectively. He planned tactics that would lead a right-wing conservative party to power, an organization which also included right-wing socialists and conservatives.

When he was released from prison in 1948 he was forbidden by SCAP to enter public office again. However, the order was soon rescinded by SCAP as a part of its 'reverse reforms'. On the day of his release sugar magnate Aiichiro Fujiyama, who had looked after Kishi's family during his imprisonment, offered Kishi the chairmanship of one of his companies and a directorship in another. His income assured, Kishi first re-established his Kishi New Party and integrated it with his 'Association for Defense of the Fatherland' to form the 'Japan Reconstruction Federation', which included some of the former members of a pre-war conservative party and elite bureaucrats.[2] Fujiyama was appointed as the head.

He consolidated the federation with the well-established ultra-conservative Japan Democratic Party whose objective was to revise the new constitution and the security treaty to allow Japan to rearm in order to defend itself. His younger brother Eisako Sato, who would be Japan's longest serving prime minister (1964–74), was the prime minister's chief cabinet secretary and Yoshida's protégé. He urged Kishi to enter the Liberal Party. Kishi understood that in order to realize his ambitions he had to enter mainstream conservative politics and requested an introduction to Yoshida.

Prime Minister Yoshida had numerous confrontations with Kishi over ideological differences and tried to dissuade SCAP from allowing him to enter public office again. On the other hand, Yoshida recognized that Kishi's connections with business brought substantial contributions and that he had become, through his political connections, a force not to be trifled with. The instability of the political environment motivated Yoshida to invite Kishi to become a member of the Liberal Party and to run for election. Kishi won a seat in the National Diet in 1953.

Kishi, the renegade politician, made a power-play within the party by accusing Yoshida of being in the lap of the Americans and the British. He succeeded in undermining Yoshida's leadership by courting senior members of the party as well as business leaders to form a new government. Calling for a new conservative party, together with 200 politicians he created the 'New Party Formation Promotion Council'. Yoshida reacted by expelling Kishi from the Liberal Party.

In 1954, Kishi and his party joined Hatoyama and other Liberal Party members and combining a strategic alliance with the Socialists formed a separate faction, the Democratic Party. Together they won 185 of the 467 seats. The Liberal Party lost almost half of their seats in the Diet. Hatoyama succeeded Yoshida as prime minister in December 1954.[3]

Kishi saw his window of opportunity and approached Liberal Party members to merge the party with the Democratic Party faction. Hatoyama was reluctant to merge the parties because he sensed that he could lose control but Kishi's political acumen and determination paid off and the Liberal Democratic Party was established in 1955 with Kishi at the center.

Kishi achieved his goal. He had succeeded in incorporating the existing conservative parties, right-wing and more moderate factions into a single conservative political party which would dominate Japanese politics for thirty-eight years. Hatoyama suffered a stroke in 1954 but continued to serve in office until 1956 before resigning. He was replaced by Tanzan Ishibashi who, due to illness, resigned within sixty-three days. On 25 February 1957, Kishi, who was foreign minister in the Ishibashi cabinet, became Japan's 56th and 57th Prime Minister (25 February 1957–19 July 1960). Eisaku Sato served as minister of finance in his brother's cabinets.

During his three years in office, Kishi's only major defeat was his attempt to centralize police power and restore to the police some of their former authority such as the right to search suspected criminals. Public protests erupted against what was regarded as the resuscitation of wartime 'thought control'. There were strikes and demonstrations supported by labor unions. Socialist law-makers rioted in the Diet and tried to kidnap the Speaker to prevent a vote. Three members of his cabinet resigned, forcing Kishi to shelve the bill.

Kishi's major success but, also, the cause of his resignation in 1960 was the revision of the US–Japan Security Treaty, which nationalists considered to favor the United States. Although the Self-Defense Force (SDF) was formed in 1954 the nationalists still wanted Article 9 in the constitution revised. The negotiation did not include the revision of Article 9 but Kishi convinced the Americans to agree to a revision of the terms of the treaty, including the curtailment of America's right to quell domestic disturbances and giving the Japanese more control over their defense. The treaty committed the United States to help Japan defend itself from attack while Japan provided bases and ports for the US military. 'Mutual defense' became central to Japan's national defense policy.

Thousands of Japanese were opposed to a military alliance with the United States and angry crowds of demonstrators surrounded the Diet buildings where Kishi and his brother Eisaku were forced to remain until they were secretly escorted by police to Kishi's residence. The LDP passed the revised treaty at a special midnight session when only a minority of Socialists were present. However, President Dwight D. Eisenhower's state visit to celebrate the signing of the revised Security Treaty was canceled due to the public protests. Kishi resigned from office and was replaced by Hayato Ikeda.

The '1955 system': the moneymen, Nobusuke Kishi/Kakuei Tanaka (1950–76)

Nobusuke Kishi

Kishi shrewdly manipulated America's intense fear of communism in the United States (1947 Truman Doctrine) and communist expansion in Asia to push his ultra-conservative party to power. America was instrumental in guaranteeing that LDP politicians would dominate the National Diet in the first election. Communist-backed labor union protests were suppressed and the CIA funneled money to the LDP in the 1950s and 1960s in order to ensure a strong, conservative government. The CIA had created a fund, known as the 'M-Fund', which was organized from the sale of Japanese military supplies left over from the war. The US State Department announced in July 2006 that, although these payments were stopped by President Lyndon Johnson in 1964, payments for 'covert programs of propaganda and social action to encourage Japanese to reject the influence of the left' continued until 1968.[4]

Kishi's Manchurian experiences were fundamental to his methods of operation while in political office. He was masterful at amassing funds for the support of his right-wing ideologies as well as for personal gain. Before he became prime minister Kishi drew in funds from his business networks cultivated during his Manchurian days and from right-wing nationalist businessmen who used their vast wealth to support right-wing politicians.

After Kishi assumed office he had freer access to *keiretsu* money particularly the conglomerates under MITI's administrative jurisdiction, such as steel, where retired MITI officials were employed through *amakudari*. Kishi systematically organized the way that campaign contributions were collected from businesses in order to alleviate the scuffle for contributions by LDP politicians belonging to different factions within the party who had been competing for funds from the same businesses.

Two of Kishi's devoted financiers were his cell-mates in Sugamo prison. Yoshio Kodama and Ryoichi Sasakawa, two wealthy right-wing nationalist businessmen, played prominent roles in Japan's post-war ultra-conservative politics. Kodama (1911–84) was a well-known underworld figure who first made his fortune in Manchuria smuggling drugs to Japan together with supplies for the military. He also engaged in espionage activities for the military. With wealth estimated to be $175 million, he was one of the richest men in Asia.

Kodama was imprisoned by SCAP as a Class A war criminal but his money, his right-wing politics and Yakuza gangster networks made him invaluable to US intelligence organizations and he was released to quell labor strikes and silence communist sympathizers. Kodama quashed all things he regarded as remotely communist and consistently supported the right-wing LDP.

Kodama met business magnate Sasakawa when he entered his fascist political party in 1931. Sasakawa, whose pre-war fortune came from rice speculation, had also been involved in numerous plots to assassinate government leaders and was jailed for several years until 1938. His was also imprisoned by SCAP for his involvement in the war effort. After his release, he built his fortune primarily through the black market and legalized gambling. In 1951, he introduced to the Yoshida government the Motorboat Racing Law to allow motorboat gambling which gave him control over all of the revenues. Kishi helped Sasakawa establish himself as the chairman of the Japan Shipbuilding Foundation, which distributed the funds to local governments where the races were held and to the Motorboat Racing Association (Kodama was appointed head of the Tokyo Motorboat Association). Sasakawa created a number of charitable foundations in his family's name, accruing considerable wealth from the activities.

Kishi was practiced at ritualizing corruption. He used public resources to produce work for businesses that supported the LDP while simultaneously reaping kickbacks for himself and members of his faction. The use of public resources was regarded as an innovative way of raising political funds that were difficult to trace. It was also an effective route for sourcing political funds connecting the LDP to the private sector. Kishi established the Overseas Economic Cooperation Fund to facilitate repayment of losses experienced by Japanese firms operating in Asia before the end of the war. The fund was designed to aid developing countries but the distribution was determined by the countries' officials receiving the aid, LDP politicians and conservative Japanese businesses.

Kishi's finesse for manipulating public funds and money laundering was legendary in the LDP. However, he was never implicated in any of the corruption scandals that involved LDP politicians after he left office.

Kakuei Tanaka: the godfather of pork-barrel patronage

Kakuei Tanaka, Japan's prime minister from 1972 to 1974, is considered the strongman-strategist in Japan's post-war politics and a key organizer of the LDP political machine. After entering the LDP in 1955, he quickly established himself as the godfather of pork-barrel politics. Tanaka was the antithesis of Kishi not only concerning his blatant and often crude methods of raising political funds but also in regards to his political career which was peppered with corruption scandals, one of which ended his term as prime minister and ultimately his political career and the Tanaka faction.

Deliberate in his methods of collecting political contributions while simultaneously lining his own pockets, he created a well-oiled political faction in the LDP, which survived until the late 1980s.

Born in Niigata Prefecture in relative poverty, Tanaka left middle school to work as an errand-boy in a construction company. The director of a local firm, impressed with the boy's ambition, financed a drafting office in Tokyo where Tanaka spent two years before entering the army in 1939. He was stationed in Manchuria for two years but contracted pneumonia and was sent back to Tokyo where he remained until the end of the war.

Tanaka's raw business sense and willingness to take risks were evident before he entered politics. In 1942, Tanaka married a divorcee seven years his senior whose father owned a medium-size construction company. When his father-in-law died, Tanaka inherited the company, renaming it Tanaka Civil Engineering and Construction Industries. Directly before the end of the war, he sold ¥15 billion ($78 million) in government bonds in Seoul.

Tanaka entered politics when SCAP planned the first post-war election in Japan in 1945. By contributing a generous donation to the Japan Moderate Progressive Party (*Nihon Shinpoto*) through a business colleague connected to the party he was accepted as a member. Although he was a member of a conservative party, Tanaka knew that, as a rough-hewn outsider, he would never be accepted by conservative voters. He relied upon his wealth to campaign in Niigata for a seat in the National Diet in 1946 and courted the socialist farmers who, due to SCAP's land reform, had received the lands they had previously been renting. Tanaka pledged to bring public works projects for schools, highways and tunnels, railroads, and land reclamation projects to the region in exchange for their votes but he lost the election.

When he ran a second time in 1947, he lost again but by a far lesser margin due to his cultivation of the rural vote. Nevertheless, he remained true to his promises to bring public works projects and subsidies to Niigata during the rest of his years in politics.

His organization in Niigata, the Etsuzankai (Niigata Mountain Association) was established during this time to serve initially as a local chamber of commerce. In the 1950s Tanaka brought the members to Tokyo by bus. He would meet with them personally and take them on guided tours of the National Diet. In 1961, Etsuzankai became a formal political organization and as Tanaka's personal association it was his strongest support group. By the 1980s there were 313 district organizations which included the majority of construction and transport companies as contributors. Tanaka himself owned several of the companies. The secretary of Etsuzankai was referred to as Niigata's 'shadow governor'.[5] At its peak, the association boasted over 100,000 members and the tightly knit group perpetuated Tanaka's image as the quintessential political boss.

After becoming a member of the Democratic Party, he finally won a seat in the Diet later in 1947. He defected to the Liberal Party in 1948 and was appointed as Vice-Minister of Justice in the Yoshida cabinet. However, later that year in December Tanaka was arrested and convicted on charges of accepting ¥1 million ($128,000) in bribes from a coal mining company in Kyushu.

After only a month behind bars, Tanaka announced his candidacy for the 1949 general election. But Yoshida and the Liberal Party had cut ties with him and refused to fund his campaign. Unfazed, he posted bail in January and was re-elected. After agreeing with Chief Cabinet Secretary Eisaku Sato to resign his post of Minister of Justice he was reinstated in the Liberal Party. Despite his new position, Tanaka was found guilty of accepting the bribe. However, he quickly filed an appeal, enabling him to buy enough time to purchase the bankrupt Nagaoka Railway that connected Tokyo to Niigata and to resuscitate its operation by 1951. He won a landslide victory that year, his campaign supported by the railway's workers as well as by a billionaire businessman who continued funding Tanaka until the end of his political career.

Tanaka served in Kishi's first cabinet as minister of posts and telecommunications. His close relationship with the future Prime Minister Eisaku Sato and his stepdaughter's marriage to Hayato Ikeda's nephew provided him with the necessary connections to two important leaders in the LDP government. In Ikeda's cabinet he served as the chairman of the Policy Affairs Council and as finance minister. Tanaka was tapped to become the LDP secretary-general when Sato became prime minister in 1965 but, again, a corruption scandal involving him in real estate deals in Tokyo resulted in Takeo Fukuda, Japan's future 67th prime minister, assuming the position. However, Tanaka was appointed the MITI minister and was popular among the electorate because of his tough stance during negotiations with the United States Trade Representative (USTR) who was pressuring Japan to accept limits on certain exports in order to decrease Japan's trade surplus with the United States.

Tanaka was the consummate user of public funds (FILP) directed towards public works projects. His close relationship with the construction and real estate industries provided avenues for bringing both substantial contributions for his political faction and huge personal wealth. In 1953, he devised the Volatile Oil Tax Law, which was the first in a series of tax-related laws, specifying the tax revenues that would be used for road construction, which developed into the biggest public works expenditure after the 1964 Tokyo Olympics. In 1966, the LDP-dominated Diet passed the Public Demand Law, which called for both central and local government to increase distribution of procurement contracts to small and medium-size firms.

Construction and real estate companies in Tanaka's camp earned enormous profits from public works projects which included railroads, schools and land reclamation. In 1972, directly before Eisaku Sato's resignation from office, Tanaka published the book *Plan for the Remodeling of the Japanese Archipelago* which pointed to crowded cities and pollution associated with rapid industrialization. Tanaka's plan called for connecting rural areas with the large cities so that Japanese could move to cleaner and healthier environments in the countryside. Tanaka's book served to boost his popularity among voters and, as importantly, it also presented him with the opportunity to profit considerably from future infrastructure projects and real estate transactions.

Tanaka achieved his ambition to become prime minister in 1972 when Prime Minister Sato resigned after serving six years in office. During his term as prime minister he took more control over the Ministries of Construction and Transportation. All of the members in Tanaka's 1972 cabinet were to be Japan's future prime ministers. Takeo Miki (1974–6) was deputy prime minister, Takeo Fukuda (1976–8) was minister of finance, Masayoshi Ohira (1978–80) was minister of foreign affairs and Yasujiro Nakasone (1982–7) was MITI minister.

Tanaka left the office of prime minister in December 1974 amidst yet another scandal. When the press criticized his business dealings with land and real estate companies the Diet held a public inquiry. The committee called the secretary of the 100,000-member-strong Etsuzankai with whom he was having an affair as a witness and Tanaka resigned rather than have her take the stand. Two years later in 1976 he was pursued by the well-documented Lockheed bribery scandal which should have ended his career in politics but, instead, the event revealed the unwavering loyalty of his constituents in Niigata who stood by him throughout the investigation.

Takeo Miki stepped in as prime minister on a 'clean government' platform and was supported by the Tanaka faction. When Tanaka was arrested on 26 July 1976 for allegedly accepting $3 million in bribes from the Lockheed Corporation for convincing ANA to order passenger planes from Lockheed instead of from McDonnell Douglas, Miki refused to intervene and thoroughly investigated the accusations. The case went to trial. Lockheed officials also implicated Eisaku Sato as well as Kodama in separate bribery cases.

Tanaka was released on ¥200 million ($690,000) bail in August. He was convicted not on bribery charges but for violating foreign exchange control laws and was given a four-year prison sentence. Tanaka remained free on appeal until his death in 1993. Nevertheless, Tanaka and his faction continued to receive substantial popular support from the electorate and Tanaka kept his seat in the Diet representing his Niigata constituency until 1985 when he and his faction finally lost power to Nakasone.

In 1990, he accepted a bribe of ¥500 million ($3.5 million) in cash reportedly delivered to his residence on a cart by Sagawa Kyubin (Sagawa Express), a trucking firm. Tanaka, in typical style, managed to persuade the court to allow him to pay a fine rather than be imprisoned. However, this scandal was the final blow to a career that set the standards for pork-barrel politics in Japan.

Shin Kanemaru: post-Tanaka moneyman

Shin Kanemaru inherited the title of 'Kingpin of the LDP' after Tanaka resigned. Not only did his close connections to the construction industry and his practices in collecting campaign funds while lining his pockets resemble Tanaka's but his fall from power also was caused by a bribery scandal involving Sagawa Kyubin.

Kanemaru was a member of the Tanaka faction and served as an aide and fund-raiser. When Tanaka resigned and was further weakened by the Lockheed bribery scandal Kanemaru, together with the future prime minister Noboru Takeshita, who also belonged to Tanaka's faction and who also was a key fund-raiser, left to form their own faction in 1985.

Kanemaru and Takeshita had known each other since 1958 when they won seats on the LDP ticket in the Lower House. Their backgrounds were similar because their prefectures, Yamanashi and Shimane, were rural and agrarian. Their fathers were wealthy sake brewers and leaders in local government. Sake brewers often engage in the rice wholesale business and, therefore, are in a position to develop good networks with both local residents and farmers. Takeshita's eldest daughter married Kanemaru's eldest son.

Kanemaru served in the army in Manchuria. When he returned to his home in Yamanashi Prefecture he built a successful liquor and concrete business in Yamanashi before entering politics. Kanemaru's backroom politics and his ability to bring in large sums of money, courtesy of his links to industry and the underworld, made him a very potent leader in the LDP and the Takeshita faction the wealthiest and largest in the LDP. Kanemaru worked tirelessly to install Takeshita as prime minister and succeeded when Takeshita was elected to office in 1988.

Kanemaru was deputy prime minister during Takeshita's eighteen-month term in office and vice-president of the LDP. As was also the case for the Tanaka faction, construction companies who wanted to expand their operations (e.g. public works projects) contributed the major share of political contributions to the Kanemaru/Takeshita faction. However, the LDP major domo was enveloped in a series of corruption scandals that ended his political career. In October 1992, he publicly admitted that he had received $4.1 million in illegal campaign contributions from Sagawa Kyubin. However, he avoided public questioning through behind-the-scenes politicking and escaped prosecution with a small fine.

In March 1993, he and a secretary were arrested for tax evasion. When public prosecutors raided Kanemaru's residence they discovered in his safe-deposit box gold bars and $50 million in cash and securities. The public outcry and the continuing media attention on the collusion scandals involving politicians, bureaucrats and big business forced Kanemaru to relinquish his seat in the Diet and tender his resignation as LDP vice-president. Takeshita had previously resigned in 1989 because of the Recruit scandal, a huge insider-trading stock scandal in 1988 that led to the resignation of many prominent politicians in all of the major parties.

Although Kanemaru was indicted for tax evasion he was not imprisoned because of his age and his health. But the string of scandals involving key LDP politicians temporarily ended the LDP's thirty-eight-year control over post-war politics.

Right-wing ultra-conservative politics: in the family domain

Since the late 1980s, a breed of young neo-nationalist politicians who did not experience the war have become a force in Japanese politics.

Furthermore, since 2001, with the exception of Naoto Kan, all of the prime ministers were groomed to be career politicians, all of them inherited their seats in the Diet from their fathers,[6] and all of them can be considered right-wing reformists. Also, three were the progeny of post-war prime ministers.

Junichiro Koizumi

Koizumi (2001–6) was a flamboyant and popular right-wing politician. His grandfather was minister of posts and telecommunications in pre-war prime ministers' cabinets and his father Junya was the director-general of the Japan Defense Agency. Koizumi entered politics to stand in the 1969 general elections. Although he was defeated, he was hired as a secretary by Takeo Fukuda, who was minister of finance in Prime Minister Sato's cabinet. When Koizumi won a seat in the Lower House in 1976, he joined Fukuda's faction. Koizumi served in 1979 as minister of health and welfare in Prime Minister Noboru Takeshita's cabinet, minister of posts and telecommunications in the Miyazawa cabinet and again as minister of health and welfare (1996–8) in Prime Minister Hashimoto's cabinet.

Koizumi will not be remembered best as a proponent of institutional reforms but for his six pilgrimages to Yasukuni Shrine, where not only Japan's military dead are interred but also Class A war criminals. The Shinto shrine was founded in 1869 and run by the military until the end of the Second World War, when the United States occupation outlawed Shintoism. Infuriated by Koizumi's apparent glorification of Japan's military past and disregard of the suffering incurred by the Chinese and Koreans during Japan's wartime occupation, China and North and South Korea lodged vigorous protests. Despite angering important trading partners, as well as business owners whose ventures in China and South Korea were vandalized by Chinese and South Korean protesters and who were extremely concerned about the ramifications for future Japanese–Chinese economic relations, Koizumi continued his visits to the shrine, ending his term as prime minister with a sixth and final visit in August 2006. He also sent SDF troops to support the United States in Iraq, which was the first time since the war that Japanese military personnel engaged in a foreign war zone.

Koizumi's nationalistic stance fueled existing tensions between Japan and South Korea. During his administration Japan reiterated its territorial rights to Dokdo, the islands that have been in South Korean control since 1953. Even though the islands have no fresh water and are primarily rock, the rich fishing grounds are highly valued. On 18 March 2005 the Japanese government declared a new national holiday commemorating the centenary of the inclusion of Dokdo as a colonial territory. In protest, South Koreans burned the Japanese flag, cut off fingers, set themselves on fire and one protester committed suicide. The dispute over the islands, known by the Japanese as Takeshima, has continued for centuries and still remains a contentious issue. On 14 July 2011, in response to a test flight by Korean Airlines of a new jumbo jet over Dokdo, Japan's Ministry of Foreign Affairs banned travel by its diplomats on the airline for one month.

Koizumi's administration drew up an education bill in June 2006 that for the first time revised the education bill that had been written in 1947. The new bill included the promotion of 'patriotism' as part of compulsory education, which had been mentioned in the 1947 bill. Although members of the opposition parties voted against the bill because it might promote nationalism, it was passed in January 2007. For the first time since the Second World War the Basic Education Law requires singing the national anthem and saluting the Japanese flag at school ceremonies (which had legally been sanctioned in 1999) as a part of children's education. The law also gives the education minister more power over local education boards and requires that teachers renew their licenses every ten years.

On 23 May 2008, the government again defied opposition to the promotion of nationalism among schoolchildren by voiding a 1949 ban on public school trips to Yasukuni Shrine. The reason given for restoring the excursions was that they would serve to educate students about Japanese culture and history.

Shinzo Abe

Shinzo Abe (2006–7) was hand-picked by Koizumi to succeed him in office. Abe was raised from birth in the ultra-conservative nationalist environment of his maternal grandfather Nobusuke Kishi, his uncle Eisaku Sato and his father, senior LDP politician Shintaro Abe.

Koizumi was very concerned about North Korea's nuclear capabilities and the unresolved issue regarding the abduction by the North Korean government of Japanese civilians. Abe was the chief negotiator for the release of Japanese who were abducted by North Koreans from 1977–83. Although the North Koreans admitted to abducting thirteen Japanese, there was conjecture that the number may have been as high as seventy or eighty and that some were abducted because they inadvertently witnessed North Korean agents operating in Japan. Abe, who was serving as Koizumi's chief cabinet secretary, accompanied Koizumi to North Korea in September 2002 to meet with North Korean President Kim Jong-il to try to normalize Japan–North Korean relations and negotiate an anti-nuclear proliferation agreement. At the meeting Kim told Koizumi that the government had kidnapped Japanese citizens to train as spies. The admission inflamed anti-North Korean sentiment among the majority of Japanese.

In August 2003, just forty-eight hours before Japan was scheduled to participate in six-way talks between China, the United States, South Korea and Russia to discuss North Korea's nuclear weapons program, angry right-wing protesters demonstrated at Niigata Port where the North Korean ferry *Mangyonbong 92* docked regularly to carry ethnic North Koreans to Wonson. The boat entered the harbor blaring North Korean patriotic songs and the Koreans returning to Japan waved North Korean flags. The Japanese demonstrators, accompanied by black armored vans spouting anti-Korean rhetoric through loudspeakers, protested that the boat smuggled drugs and counterfeit money into Japan.

Many North Koreans immigrated to Japan during and directly after the war and ethnic North Koreans commuted regularly to visit relatives. Since Japan did not have formal relations with North Korea, the ferry service enabled direct contact with their families. It took six months until the ferry service was restored. However, the service was suspended again in 2006 as part of economic sanctions against Korea after North Korea fired a missile over Japan.

Shortly before Abe assumed office he published the book *Toward a Beautiful Country*. The best-seller promoted his right-wing ideology and his intent to perpetuate national pride through the reconsideration of the chronicling of events of Japan's engagement in Asia during the Second World War, post-war period and the revision of the pacifist constitution in order to sanction Japanese military participation in peacekeeping operations abroad. Abe argued that the reporting of events regarding his grandfather's revision of the US–Japan Security Treaty in 1960 was misleading and that the initial 1951 treaty inhibited the Japanese from prosecuting Americans who perpetrated crimes in Japan (e.g. military personnel). He claimed that his grandfather negotiated a revised treaty that gave Japan more autonomy from the United States and gave the Japanese more power to prosecute.

Abe insisted that the Japanese had been a conservative nation for several thousand years, that their society was a conservative system and that the word 'conservatism' was regarded negatively while the term 'liberalism' was considered positively. He took refuge in the fact that in recent years, the number of conservative journalists and commentators who espoused nationalism was on the increase.

During his administration Abe promoted bills that would serve the national interests, such as the reinstatement of the Ministry of Defense and the revision of Article 9 in the Constitution to allow the SDF to protect Japan if Japan was attacked. Abe maintained that the history reported in school textbooks did not reflect a realistic picture of Japan's wartime engagement in Asia.

In 2001, Abe supported the Ministry of Education, Culture, Sports, Science and Technology's (MEXT) approval of the *New History Textbook* for high-school students, which glosses over the wartime atrocities committed by the Japanese military in Asia, such as the abduction of comfort women. The textbook refers to the war in the Pacific as 'The Greater East Asia War', the term used by the Japanese government after Japan's bombing of Pearl Harbor in 1941 and by Japanese nationalists. Other proponents of the textbook included former METI vice-minister Keiji Furuya and Shintaro Ishihara who won a fourth term as Tokyo's governor in April 2011. *Sankei Shimbun*, a conservative daily which is owned by the same ultra-conservative media group that also owns the textbook's publisher, ran a series of editorials supporting the textbook. The book's authors were academics who were members of the right-wing Society of Textbook Reform and who believed that Japan had entered Asia to liberate the region from the control of white colonists and that the so-called atrocities were merely 'normal excesses' committed by all armies. Koizumi's administration also advocated the textbook and admitted to organizing town meetings and paying educators to give rehearsed speeches favoring the textbook.

Despite China's and South Korea's anger over the manipulation of facts, negative reactions from the foreign press, heated debate in the Diet and strong resistance in local governments, by 2004 the book had been adopted by a number of prefectures. In April 2004, the Tokyo Metropolitan Board of Education officially adopted the textbook, thanks to the persistence of Governor Ishihara. In 2009, Yokohama, with a population of 3.7 million, became the first city to adopt the controversial textbook. In 2011, the Yokohama educational board approved further use by 27,000 students in 150 middle schools from April 2012.

For several years before the Basic Education Bill had been formally drawn up teachers who refused to press students to sing the anthem and salute the flag were reprimanded. In May 2004, the Tokyo Board of Education reprimanded 171 teachers for refusing to comply with the law. By 2007, 450 teachers had been punished by the board for non-compliance through suspension or the withholding of wages, actions which resurrected memories of the Second World War and the era of a fascist Big Brother state and the intensification of government controls over society.

On 15 December 2006, a law was passed in the National Diet for the formal upgrading of the Defense Agency to a Ministry of Defense (MOD). The re-establishment of the MOD reflected the government's concerns over North Korea's missile and nuclear capability but it was also specific to promoting patriotism and attempts to revise the constitution. On 9 January 2007, at the inauguration ceremony Shinzo Abe and his minister of defense Fumio Kyuma reviewed the Japanese honor guard. Abe proclaimed that self-defense was a nation's inalienable sovereign right and that as prime minister he was proud to open an institution responsible for national defense.

Abe's career as prime minister was short-lived due to the ongoing pension debacle, a series of financial scandals involving three cabinet ministers (the minister of agriculture committed suicide) and a gaffe made by Fumio Kyuma at an address given at the university in Chiba Prefecture. Kyuma stated that he felt that the dropping of the atomic bomb on Nagasaki by the US was unavoidable and probably served to save thousands of Japanese lives and to deter the Soviet Union's invasion of Japan. Although Kyuma gave a public apology and resigned his post, the electorate was thoroughly disgusted. The LDP loss of the majority in the Upper House election to opposition parties in August forced Abe's sudden resignation after only a year in office.

Yasuo Fukuda

Yasuo Fukuda also resigned after only one year in office (2007–8). He was Japan's first prime minister whose father Takeo had been a prime minister and he also assumed the office at the same age as his father at seventy-one. Takeo Fukuda (1976–8) succeeded Takeo Miki whose 'clean government' power base proved too weak to carry him on to a second term. After graduating from Tokyo Imperial University, Fukuda entered the Ministry of Finance and served in top positions in economic planning and as chief cabinet secretary during the Second World War. At the beginning of Japan's post-war era he was the director of the General Accounting Office. Fukuda entered politics in 1952 when he won a seat in the Lower House as a member of Yoshida's Liberal Party. He served in successive prime ministers' cabinets as minister of agriculture, finance and economic planning.

Fukuda fully expected to succeed Eisaku Sato, who was Japan's longest serving prime minister (1964–72), but Tanaka, who was Sato's MITI minister, proved too popular and won the LDP presidency. Fukuda and Tanaka served together as ministers in various cabinets but there was a significant difference in the way they operated in the political arena. Fukuda considered Tanaka, the heavy drinking and smoking back-room politician, as uneducated, uncouth and corrupt. During his administration Fukuda tried to end factionalism in the LDP by establishing primary elections for party president. Ironically, he lost the first election to Masayoshi Ohira and was forced to resign. Takeo was grooming his eldest son for political office and Yasuo, his second son, was not expected to enter the world of politics.

Yasuo worked in an oil company until he was forty until the untimely death of his brother whom he succeeded as his father's political secretary. However, he remained on the periphery of politics until 1990 when he won a seat on the LDP ticket in the Lower House as a representative from Gunma Prefecture (his father's former seat). Fukuda was elected deputy director of the LDP in 1997 and became chief cabinet secretary from 2000–4. When the investigation of the SIA pension scandal revealed that he had missed premium payments he resigned the post.

Compared to Koizumi and Abe, Fukuda could be considered a dove in his policies regarding Japan's relationship with China and South Korea. He recognized the importance of maintaining strong economic cooperation and tried to terminate visits to Yasukuni Shrine. Nevertheless, a bribery scandal involving the vice-minister of the newly established MOD and the failure to resolve the pension issues and other domestic problems reflected his weak leadership and he resigned suddenly to leave the prime minister's post open to Taro Aso, who had entered the election for LDP president in 2007 but was defeated by Fukuda.

Taro Aso

Taro Aso (2008–9) had no better luck than his two predecessors in the Executive Office. By Japanese political standards Aso's birthright was sterling. His great-great-grandfather was Toshimichi Okubo, his maternal grandfather was Prime Minister Shigeru Yoshida, his wife was the third daughter of Prime Minister Zenko Suzuki (1980–2), and one of his sisters was the wife of the first cousin of Emperor Akihito. His father was the chairman of the Aso Cement Company and a close colleague of Kakuei Tanaka. Aso's education was also elite. He studied at Stanford University and later at the London School of Economics. He joined the Aso Mining Company and served as president for four years before entering politics in 1979. He ran for a seat in the Lower House as a representative from Fukuoka Prefecture and won, leading the way to successive victories.

Aso was minister of internal affairs, posts and telecommunications in Koizumi's 2003 cabinet. He became minister of foreign affairs in 2005. When Koizumi resigned Aso put his hat into the arena. His platform regarding foreign relations was 'Arc of freedom and prosperity: Japan's Expanding Diplomatic Horizons'. Aso lost the election to Abe, who was considered a 'dove' compared to Aso. When he ran a second time for the LDP presidency, Fukuda won the election and succeeded Abe as prime minister. Fukuda appointed Aso as secretary-general of the LDP, which put him into the position of second-in-command. When Fukuda resigned Aso was chosen to succeed him.

Aso's right-wing ultra-conservative opinions were well publicized in the press during his political career. In 2005, he touted Japan as a monoculture with one race, one language and one civilization, excluding other ethnic groups such as Chinese and Koreans.

In 2006, Aso admonished the Japanese that China, with a population of one billion people, possessed nuclear bombs and had expanded its military budget by double figures for seventeen consecutive years. He also praised his forefathers for implementing compulsory education in Taiwan during its occupation.

In 2008, the Western media reported that Aso Mining had forced 300 Allied prisoners of war to work without pay in the Aso Mining Company in 1945 and that two Australians had died. Furthermore, 10,000 Koreans were recruited to work in the mines under cruel conditions, many of whom also died. Subsequently, the Ministry of Foreign Affairs investigated the charges and in December confirmed that two Australians had died. Although former laborers requested an apology, Aso refused to reply.

His fiscal policies hinged on the release of several stimulus packages, which did little to bring the economy out of deflation. Aso's flamboyant behavior of wining and dining nightly at expensive bars and restaurants did not help to endear him to voters. The LDP was demolished at the polls by the DPJ in the Tokyo Metropolitan election in July 2009. Despite opposition from LDP leaders, Aso called a snap election in August. He reasoned that the fund-raising scandal in May regarding DPJ kingpin Ichiro Ozawa and a construction company, which forced Ozawa's resignation as the party's president, would provide enough of an incentive for voters to remain loyal to the LDP. He was wrong. The DPJ decimated the LDP in the Lower House election on 31 August.

Yukio Hatoyama, the president of the DPJ, became Japan's first DPJ prime minister on 9 September 2009.

Yukio Hatoyama

Yukio Hatoyama (September 2009–April 2010) entered office with a political lineage equal to Aso's. His grandfather was Prime Minister Ichiro Hatoyama and his father Iichiro was a former MOF vice-minister (1971–2) and the minister of foreign affairs in Takeo Fukuda's cabinet. Yukio's mother was the Bridgestone Tire heiress and her two sons Yukio and Kunio were said to be worth over $100 million. Hatoyama received an engineering degree from Tokyo University and a doctorate in engineering from Stanford University. After teaching a management course at Senshu University in Tokyo he worked as his father's private secretary in 1983. In 1986, as a member of the LDP he took over his father's seat in the Diet. Hatoyama left the LDP to form the New Progressive Party with Naoto Kan and later the DPJ in 1998.

Hatoyama, who fended off criticism that his mother funded his political campaigns, described the DPJ as a people's party and stated that because the LDP had been in power for so long, the Japanese had forgotten how to express discontent and 'people power'. At the beginning of his administration, he, along with the international media, pronounced the election as 'revolutionary' and claimed that it would end the leadership of the bureaucrats. However, his term in office was cut short by the same problems that his predecessors had failed to resolve and the issues were becoming more complex.

Political grid-lock paralyzed policy-making decisions and Hatoyama's fiscal policies that included cuts in government spending tackled only 25 percent of the actual target of $32.8 billion. The budget was stuffed with fiscal stimulus packages but failed to raise employment figures and consumption above 50 percent (domestic consumption in Japan accounts for almost 60 percent of GDP). Hatoyama's indecisiveness regarding the relocation of Futenma military base, a string of political corruption scandals, including an investigation into a campaign funding scandal involving a $4 million donation that Hatoyama had not properly reported, and a second political funding scandal implicating Ichiro Ozawa, the vice-president of the DPJ who refused to resign from the DPJ, caused Hatoyama's fall from grace within eight months after taking office. Hatoyama later told prosecutors that perhaps the funding for his campaign attributed to his family was disguised as 'donations'. When he resigned in April 2010 Naoto Kan, who was Hatoyama's finance minister and the deputy prime minister, took the helm. Hatoyama continued to collaborate with Ozawa on party issues and hoped that after Kan resigned from office he would regain a key position in the DPJ.

Ichiro Ozawa: protégé of Tanaka and Kanemaru

Ozawa represented Japan's mainstream ultra-conservative political environment during Japan's post-war period that evolved through interpersonal connections between politicians and businesses to become a tangled web of vested interests, collusion and corrupt practices.

Ozawa is credited with engineering the 2009 landslide victory of the DPJ. Known as the 'shadow shogun' he was regarded as a key political strategist, back-room deal-maker and a fund-raiser in the same mode as Tanaka, his mentor. Shin Kanemaru was also a mentor. Tanaka taught Ozawa the importance of establishing strong relationships with smaller firms that were trying to compete with big established companies for a share of tight markets. He also taught Ozawa that a close relationship with the construction industry was essential to a politician's purse.

Ozawa was born in the rural and agrarian Iwate Prefecture in northeast Japan. His father, who owned a construction company, held a seat in the Lower House. When he died, Ozawa stood for election in 1969 as a member of the LDP and won. He joined Tanaka's faction and rapidly gained popularity as did future prime ministers Tsutomo Hata and Ryutaro Hashimoto with whom he jostled for position in the Tanaka faction. He impressed LDP leaders with his skill at manipulating opposition parties to pass unpopular bills in the Diet but because of his unfettered ambition, and dictatorial methods, he was also disliked by many in the LDP. A right-wing reformer, he transferred to the Takeshita–Kanemaru faction after Tanaka left office. When Kanemaru was arrested in October 1992 for the Sagawa Kyubin affair, Ozawa was also interrogated about escorting him to meetings with Sagawa management. His retort to prosecutors was that he was merely going along to act as barman. Ozawa also had been implicated in the Recruit scandal in 1988 for his involvement in insider-trading transactions.

In 1993, he defected from the LDP to start the small right-wing reform Renewal Party (*Shinseito*) with Hata and helped to form an eight-party coalition in Prime Minister Hosokawa's government. Ozawa's key objective was to reform the election system that would encourage fairer representation in the Lower House and result in a true two-party system. It would also serve to eliminate some of the collusion scandals that had been dogging the LDP. By coincidence, in December 1992 when he was still an LDP member Ozawa was implicated in a money scandal which involved Kajima Construction, one of the big six general contracting companies that had been charged with bribing numerous politicians in order to win contracts. Ozawa called a press conference to deny reports by the *Asahi Shimbun* that the contributions from Kajima were illegal. The coalition government passed the reform bill in 1994, which fourteen years later served to propel the DPJ into the majority in the Lower House elections.

Ozawa also wanted the reform of the bureaucracy to allow politicians and bureaucrats to share equally in policy planning. He was determined to bust the power base of the bureaucracy and tried to entice elite bureaucrats to leave their ministries. There is speculation that he was involved with Hiroshi Kumagai, the MITI minister in Hosokawa's cabinet, in the dismissal of Masahisa Naito. Nevertheless, few bureaucrats took Ozawa's efforts to poke holes in their system seriously.

Hosokawa was also accused of receiving political funds from the Sagawa Kyubin and resigned in April 1994. Hata succeeded him as prime minister but held on for only three months because of Ozawa's habit of alienating colleagues by pressuring them to accept policies he vigorously promoted. After failing to convince the Socialists, whose party was a member of the coalition, to agree to certain issues, they left Hata's coalition and the LDP regained seats in the Lower House general elections in June, forcing Hata's resignation in June.

Ozawa plowed ahead to form the Pioneers Party (*Sakegaki*) in 1995 and merged it with the Renewal Party. He defeated Hata in the election for secretary-general of the party but it was a meaningless exercise because Hata and the majority of members left the party to form other small parties. Ozawa went on to form the Liberal Party, which participated in the LDP coalition government. In 2003, Ozawa merged the Liberal Party with the DPJ. He became the DPJ secretary-general but his imperiousness led to confrontations with party members, especially with Naoto Kan.

He was elected as president of the DPJ in 2006 and his campaign tactics led the party to its biggest victory in the Upper House. He resigned in 2007 in protest after Prime Minister Fukuda proposed a coalition between the LDP and the DPJ but he was persuaded by DPJ leaders to remain in the post. He won the presidency three consecutive times. In May 2009, Ozawa was pressured to resign due to his alleged involvement in the political funding scandal. However, when Ozawa masterminded the DPJ victory in the Lower House, Prime Minister Hatoyama appointed him secretary-general of the party in 2009.

Ozawa was taught by his mentors how to garner the support from local electorates and grass-roots politics. He creatively cultivated a devout following of young supporters throughout the 1990s. Many of his lieutenants were first hired as staff to perform menial chores in his home such as cleaning and walking his dogs. If they proved to be hard-working and loyal, they were promoted to secretarial positions in his Tokyo office.

The loyalty of two lieutenants and their willingness to support him regardless of the methods involved resulted in the arrest on 3 March 2009 of Ozawa's chief political secretary Takanori Okubo for allegedly accepting illegally $250,000 in political contributions from Nishimatsu, a construction company. Private firms are allowed to contribute to political parties but not to individual politicians. The money was passed through Ritsuzankai, Ozawa's fund body. Ozawa claimed ignorance of the contribution. Nishimatsu was already being investigated for other corruption scandals, which may have led prosecutors to Ozawa's door. The public prosecutors pursued the case and relentlessly investigated his alleged financial irregularities. But Ozawa, ever the survivor, as the new secretary-general of the DPJ kept an indignant silence and denied requests from the Diet to appear at hearings during the rest of 2009, relying on his popularity among voters to defuse the situation. However, 75 percent of voters thought that he should resign as secretary-general after prosecutors raided Ritsuzankai again to discover $4 million in cash in illegal contributions from Kajima Corp.[7] Prosecutors also searched the Diet members' building office of a former aide, Tomohiro Ishikawa, who was a member in the Diet's Lower House and Kajima Corp's offices. Ishikawa and Okubo were also charged with not registering contributions in the Ritsuzankai books in order to hide the receipt of ¥50,000 each from Mizutani Construction related to a dam construction project in Iwate. Ishikawa had proven his loyalty to Ozawa by shoveling snow and scrubbing floors at his house and doing clerical work at Ritsuzankai before entering politics.

Ozawa brushed off accusations of any wrongdoing stating that, although he and his aides may have misjudged conditions, there was no intention to violate the law. Ishikawa, aged thirty-six, was indicted for accepting illegal contributions and resigned from the National Diet on 2 December 2010. On 31 January 2011, a special panel of lawyers who had been assigned to investigate the case returned with an indictment. Ozawa's continued resistance to face questioning in the Diet and Kan's administration's shrinking public approval rating (17 percent) forced the DPJ Standing Officers Council to officially suspend Ozawa from the DPJ on 20 July. Nonetheless, like his mentor Tanaka, despite his indictment and pending trial, Ozawa maintained his power base. His group of 120 supporters was significantly larger than the groups supporting the other

DPJ politicians. When Kan was about to resign in late August, some candidates sought Ozawa's support, suggesting that his suspension from the party be reconsidered. When Ozawa and Hatoyama announced that they were backing Banri Kaida, Kan's minister of trade and industry, the Japanese press predicted that he would win. However, Yoshihiko Noda, Kan's finance minister and key supporter won the presidency. On 26 September, Okubo and Ishikawa were convicted for making false entries in Ozawa's political funds records and received suspended prison sentences. Ozawa's power base was eroding rapidly.

The Recruit scandal: interpersonal networks of politicians, bureaucrats and big business

The massive insider trading and corruption scandal in 1988 that brought down Prime Minister Takeshita's entire administration, the resignation from office by many key politicians and the arrest and indictment of powerful businessmen is a pertinent example of collusion between cross-party ultra-conservative politicians, bureaucrats and big business and of money politics during Japan's post-war period. The Recruit scandal, involving 155 prominent figures, was regarded as the most pervasive of all time and much bigger than the Lockheed scandal. It was also credited with spurring the defection of members from the LDP to form the New Japan Party in 1992 and Morihiro Hosokawa's installation as prime minister in August 1993.

Hiromasa Ezoe established Recruit with $2000 as an advertising-sales agency in1960. Ezoe, whose motto was 'In This World Money Comes First', built the tiny company into a $3.2 billion conglomerate with subsidiaries that engaged in human resources, real estate, supercomputers, restaurant and hotel management, information-publishing, property and telecommunications. Throughout the years, in order to expand the company's businesses, Ezoe habitually offered his company's unlisted shares before they went public to politicians, bureaucrats and businessmen who made substantial gains when the shares were released on the open market.

From 1984–6 Ezoe sold shares of the real estate subsidiary Recruit Cosmos for $20 per share before it went public to politicians, bureaucrats, business leaders and journalists. When Cosmos went public and the shares were sold on the over-the-counter, asset-inflated stock market the price per share tripled and the purchasers reaped millions of dollars. It was common practice for Japanese companies to sell stock to influential buyers at reduced prices but the magnitude of the sale of more than 885,000 shares was unprecedented.

Takeshita had been in office since November 1987. The scandal hit the front pages of major dailies in June 1988. Ezoe was arrested, and during the eight-month investigation Ezoe told prosecutors that he had spread millions of dollars in cash and unlisted stock to 100 politicians, bureaucrats and businessmen. Prosecutors arrested or indicted nineteen people, among them the former chairman of Nippon Telephone and Telegraph Hisashi Shinto and two of the corporation's top executives, Kunio Takaishi, the former vice-minister of the Ministry of Education and Takaishi Kato, the former vice-minister of the Ministry of Labor.

The implication of bribery among a number of Takeshita's cabinet ministers did not help his public approval rating which had already sunk below 30 percent. Takeshita's finance minister, Kiichi Miyazawa, resigned in December 1988, claiming that an aide had purchased stock using his name. Takeshita quickly reshuffled his cabinet, replacing fifteen of his twenty ministers.

Nippon Telephone and Telegraph (NTT), a Special Corporation managed by the former Ministry of Posts and Telecommunications, was the domestic provider of telephone networks and was in the process of undergoing partial privatization. Ezoe needed strong political connections to ensure that the regulators would not break up NTT's networks into smaller private companies that would be competitive with Recruit. Offering shares was an innovative and cost-free method of promoting influence. When Miyazawa resigned in December, Hisashi Shinto also admitted his involvement in the stock deal and resigned from his post of chairman.

In 1986, Ezoe met with Shinto to arrange two business transactions with NTT. The first was getting Recruit into the lucrative market of reselling special telephone circuits to businesses. The second was NTT's procurement from Recruit in 1988 and 1989 of two US-made Cray Research supercomputers, which NTT would resell to Recruit at a reduced price. At the same time, Mr. Shinto, who was a close friend of Takeshita's predecessor Prime Minister Yasujiro Nakasone, was given 10,000 shares in Recruit Cosmos. Shinto used the shares for two NTT slush funds for direct contributions to prominent politicians and for purchases of tickets to political fund-raising events.

When Takeshita was questioned by prosecutors in October 1988 he denied any misconduct, but in April 1989 he conceded that he and members close to him had received $1,000,000 from Recruit. Although he denied making any profit from the Recruit shares he admitted that Recruit had purchased more than $570,000 worth of tickets to two fund-raising events for him in Tokyo and Iwate Prefecture (Ozawa's home turf) in May 1987. Although such contributions were legal the

Recruit contribution exceeded the limit that had been set after the Lockheed scandal.

On 25 April, Takeshita announced that he would resign from office. A former aide committed suicide in penance for his leader's humiliation. Prosecutors concluded after eight months of investigation that there was not enough evidence to indict Takeshita. Kiichi Miyazawa succeeded Takeshita in June and remained as prime minister until the LDP defeat in 1993.

On 10 October 1990, the eighty-year-old Shinto, considered one of Japan's most powerful business figures, was convicted of receiving bribes. He was fined $170,000 and sentenced to a two-year prison term but the sentence was suspended because of his age.

Other well-known figures involved were the chairmen of the *Yomiuri Shimbun* and the *Nihon Keizai Shimbun*. Yasujiro Nakasone also conceded receipt of Recruit shares and resigned from the LDP to become an 'independent'. He escaped indictment as did a number of other prominent government leaders.

Toshihiro Ono, Ezoe's chief aide, who claimed he received instruction from Ezoe to bribe Takaishi Kato and two Diet members from 1985–6, was tried and convicted in September 1997 and received a two-year suspended sentence. Ezoe was convicted of massive political bribery on 3 April 2003. He was given a three-year prison term but the sentence was suspended for five years.

The Ministry of Defense scandal: interpersonal networks of politicians, bureaucrats and big business

In October 2007, a bribery scandal involving the vice-minister of the MOD, the defense minister in Abe's cabinet and the managing director of a large trading company was widely covered by the Japanese media. The scandal is specific to Koizumi and Abe's push for the ministry's reinstatement and is also a fine representation of the consequences of interpersonal networks between ultra-conservative politicians, the bureaucracy and big business. The incident also draws attention to the ministry's procurement practices.

Japanese trading companies (in METI's administrative jurisdiction) play a major role in the procurement of defense-related equipment from foreign manufacturers because the ministries are prohibited by law to purchase equipment directly from the suppliers. The trading companies facilitate such items as contracts and customs procedures for foreign manufacturers. They also collect information for the Japanese government that identifies defense products that are scheduled to go on the market during the next five to ten years.

Although Japanese manufacturers of heavy industries (also in METI's jurisdiction), among them Mitsubishi and Kawasaki, produce weapons for domestic use, they are forbidden by law to export. Therefore, they cannot compete with foreign manufacturers who produce under licensing agreements in Japan. As a result, military hardware purchases can be ten times more expensive than in the United States and European countries.

About 100 trading companies are registered with the MOD and they are fiercely competitive about being designated as the sole agent who can monopolize procurement overseas and the facilitation of contracts: 7 percent of the annual budget for the procurement of equipment by Japan's SDF is spent on imported components and other parts that are handled by the trading companies. In fiscal 2005, the total contract amount for imports was ¥152.5 billion ($1.38 billion). Over 70 percent of the contracts awarded are said to be no-bid contracts.

In October 2007, former vice-minister of MOD Takemasa Moriya admitted to the Upper House to breaking ministry regulations when he and his wife accepted invitations to play golf with Motonobu Miyazaki, the president of Yamada Corporation, a trading company that had been receiving an unduly large number of contracts from the ministry for years. Yamada represented many large foreign military manufacturers in Japan such as Lockheed Martin, and General Electric.

Regulations forbid MOD officials and members of the SDF from playing golf with the employees of companies that conduct business with the ministry. Moriya testified that he had accepted twelve golfing trips worth ¥3.89 million ($33,143) and 108 single-day golf excursions worth ¥4.9 million ($41,748). Moriya and his wife were arrested in November for accepting bribes during 2003–6 that were estimated to exceed ¥10 million.

Moriya joined the former Japan Defense Agency in 1971 as a career official on the fast track. He was appointed vice-minister in August 2003 and played a fundamental role in the upgrading of the agency to the MOD. Moriya, who was known as 'the emperor' because of his powerful status, had emphasized to his staff the importance of ethical behavior. Moriya had already resigned from the ministry in August 2007 after a disagreement with cabinet defense minister Yuriko Koike concerning his successor. Moriya, who was known to keep detailed notes regarding whom he met and the subjects discussed, burned all of his records. He later claimed that he was angry with Koike who was pressuring him to resign.

According to Miyazaki, Moriya had been a long-time friend and stated that he frequently played golf with him. Further testimony given by

Moriya to the House of Councilors (Upper House) Committee on Foreign Affairs and Defense and to prosecutors in Tokyo District Court following his arrest for embezzlement, clearly showed that Miyazaki had entertained Moriya and his wife often, had given Mrs. Moriya gifts when her husband had been the head of the aircraft section at the former Defense Agency and that the section handled parts that Yamada sold to the agency.

When he was the director of the agency's Secretariat in 2000, Moriya opened two bank accounts under the names of two retired officials. He asked Miyazaki to deposit ¥500,000 ($4,696) into each account. The money was transferred from a Yamada Corp. bank account in Japan.

He also allegedly received ¥300,000 ($2,590) from Miyazaki as a congratulatory gift upon his promotion in 2003. Moriya was arrested again in November for allegedly taking ¥2.18 million in cash gifts in May and June 2004 ($40,560) and ¥1.45 million ($12,487) in February 2006, using the funds to pay off family debts.

Moriya had known since 2001 that Yamada's dealings were tainted with corrupt practices when the Defense Agency discovered that Yamada had overcharged the agency for parts produced by the British company BAE Systems and that it had forged BAE's documents that presented the correct quotes and submitted its own documents quoting costs. The agency duly notified BAE.

In 2002, when the agency was investigating the overcharge, Yamada hired a man to impersonate a BAE executive and coordinated a meeting with the agency officials in the United States. Later the agency received a letter ostensibly from BAE stating that Yamada should not be held accountable. However, BAE had not given permission to Yamada to submit such a letter. Yamada revised the contract and returned ¥100,000 million ($801.3 million) to the agency. General Electric temporarily canceled its agreements with Yamada.

After prosecutors raided Yamada's offices on 30 November 2007, the MOD disclosed that Yamada had engaged in 666 contracts since FY 2002, acting on behalf of foreign defense equipment producers and that 116 were accepted by the agency's head office and the other 550 were accepted by local offices. Moreover, Yamada Corp. had overcharged the ministry on a number of contracts and there were at least twelve cases of forgeries of documents.

Miyazaki was in charge of business transactions for a considerable period of time. He had wooed not only Moriya but also two other senior politicians in the LDP. According to Moriya's testimony, Miyazaki had entertained Fumio Kyuma, who had served in Hashimoto's cabinet as state minister and the head of the Defense Agency (1996–8) and in Abe's cabinet as defense minister. Moriya also named Fushiro Nukaga who was minister of the Defense Agency in the late Prime Minister Keizo Obuchi's cabinet in 1998 but resigned when he was accused of accepting funds from a public entity that supported small business development. Nukaga returned the money and Yasuo Fukuda, who was chief cabinet secretary at the time, accepted his explanation. Nukaga was reinstated in his former post, remaining in the position in Koizumi's cabinet until 2006.

Miyazaki established slush funds in a number of bank accounts that were managed by Yamada's US subsidiary, which was directed by Naoki Akiyama. After a disagreement with Yamada Corp.'s owner over management issues, Miyazaki left the company in 2006 and established the Nihon Mirise Corp. (NMC). He used the slush funds to entertain top officials in the State Department and the Pentagon in an attempt to procure business that came from the realignment of US forces in Japan, which had reached a final agreement in May 2006. Nihon Mirise stood to earn ¥10 billion ($86.1 million) in profits for contracts on US military bases. Nihon Mirise Corp. was now competing with Yamada Corp. for contracts as agents for American firms such as Northrop Grumman and General Electric.

Akiyama was the director of a Japan–US Center for Peace and Cultural Exchange, a Foreign Ministry-linked organization that was established in 1968. The members of the organization are American and Japanese politicians who lobby for defense industries and are experts on national security issues. Miyazaki, who allegedly gave ¥30 million ($258,000) to the center in 2006, was a former head. Fumio Kyuma also served as one of the directors. The Defense Agency commissioned the center to conduct research regarding a project to remove poison gas shells in Fukuoka Prefecture that remained after the war.

Akiyama was also the director of the Congressional National Security Research Group. Yamada Corp.'s president and CEO Yoshiko Yonezawa was a member of the executive board. Other members of the board included Prime Minister Fukuda, former defense ministers Fumio Kyuma and Fukushiro Nukaga, and Seiji Maehara, the former DPJ president, minister of transport and minister of foreign affairs in Prime Minister Naoto Kan's cabinet.

It was later made known that Yamada Corp., not to be out-gunned, had planned to bribe two senior US government officials to act on behalf of Yamada, which imported parts used in the Air SDF's E-2C surveillance aircraft produced by Northrop Corp. And to ensure strong interpersonal networks with government officials Yamada Corp. employed Shozo

Azuma, a member of the DPJ and former member of the Lower House, and hired through *amakudari* several retired MOD officials and their family members.

Both Moriya and Miyazaki were arrested again in January 2008 on charges of bribery. Moriya was also charged with perjury. At his trial at Tokyo District Court the following April, Moriya pleaded guilty to charges that he had accepted a total of ¥12.49 million ($120,264) from two Yamada officials from 2003–7.

The right-wing ultraconservative mind-set continues: Yoshihiko Noda (30 August 2011–)

After Noda won the DPJ presidential election on 29 August he was forced into a run-off with LDP president Sadakazu Tanigaki. Noda narrowly won the election for prime minister by three votes. Noda is the sixth prime minister in five years. He was born in Chiba Prefecture in May 1957. His father served in the Self Defense Force. After graduating with a degree in political economics from Waseda University, he attended the Matsushita Institute of Government and Management[8] and graduated in 1985. Noda began his political career at twenty-nine when he won a seat in the Chiba Assembly. In 1992, he entered the Japan New Party and the following year won a seat in the Lower House election but lost the seat in the 1996 election. Noda entered the DPJ in 1998 and in 2000 won a seat in the Lower House. He was serving as finance minister in Kan's cabinet when he stood for election for president of the DPJ on 29 August.

Noda is considered a fiscal conservative and a defense hawk and strongly supports the US–Japan Security Treaty. He may disagree with members of the LDP about the methods of the reform of Japan's political system but his ideology regarding Japan's position in the global political economy reiterates the nationalistic attitudes of LDP politicians. Although Noda, as a member of Kan's cabinet, did not visit Yasukuni Shrine on 15 August, marking the sixty-sixth anniversary of Japan's surrender in the Second World War, at a press conference held the same day he contended that Japanese Class A criminals were not criminals and that there was no reason for a prime minister not to visit the shrine. Incensed by Noda's rhetoric, the South Korean Foreign Ministry stated that his words were in denial of Japan's past history of aggression.

Noda, who has a black belt in judo, enjoys watching combat sports and drinking sake. He admires Robert Kennedy who, besides serving as US Attorney General (1961–4), was also a legal aid of Republican senator

Joseph McCarthy during the infamous televised Army–McCarthy hearings in 1954 when McCarthy tried to uncover subversive activities in government organization.

Noda also deeply admires Konosuke Matsushita (1894–1989). Matsushita, who was a devout Zen Buddhist of the Rinzai School, founded the Matsushita Industrial Electric Co. Ltd. (now Panasonic Group) in 1929 and is regarded as the guru of modern Japanese industrialists. His treatises on management, production, employee development and welfare, guiding business through crisis served as textbooks for big business. Matsushita established the PHP Institute (Peace, Happiness and Prosperity) in 1946 to restore the Japanese identity he felt was crushed by the war.

Matsushita established his Institute of Government and Management in 1979. The institute incorporates Matsushita's personal philosophies and values that were promoted by the PHP. The Institute's 'Basic Principle' encourages 'deep love of our country and our people' and the contribution to the peace, happiness and prosperity of all people. The curriculum includes Zen meditation and the Martial Arts.

Since he had backed Ozawa's suspension from the DPJ, in order to placate Ozawa's faction, Noda appointed Azuma Koshiishi, a staunch Ozawa ally, as DPJ Secretary-General. Koshiishi's efforts to bring consensus among DPJ and LDP law-makers on unpopular fiscal policies such as raising taxes were vital to Noda's platform. The prime minister retained some of Kan's former cabinet ministers to serve in his cabinet. He appointed Yukio Edano as METI minister, Koiichiro Gemba as foreign minister and Seiji Maehara as the chairman of the Policy Research Committee to whom he would give greater authority to make policy decisions. Both Gemba and Maehara graduated from the Matsushita Institute of Government and Management.

Seiji Maehara

A number of leaders in both the LDP and the DPJ have strong anti-communist sentiments and view China's economic expansion overseas and growing military budget as a signal that Japan is losing its competitive edge in outward investment in global markets and that China is fast becoming a military superpower in Asia. China was said to be spending in 2010 about 7.5 percent of its GDP which was actually less than other countries, including the United States. Naoto Kan voiced the same concerns about the rise of China's military expenditure and the lack of transparency in its military budget. At a news conference on 24 October 2010 Kan emphasized that the security around Japan was being jeopardized by North Korea's missiles and nuclear development and by China's intensification of marine activities.

The remarks concerning China were sparked by a territorial dispute over the Senkaku Islands in the East China Sea that had continued since the 1970s. Japan considers the islands as a part of Okinawa. When a Chinese fishing trawler collided with a Japanese coastguard vessel in September, the captain was arrested and held in custody for seventeen days before being released. Japan and China were also engaged in a tug-of-war over the territorial rights to massive gas reserves in the South China Sea and the incident with the trawler further inflamed existing tensions regarding a territorial dispute over the huge gas and oil reserves and bountiful fishing grounds in the East China Sea.

As the minister of foreign affairs in Kan's second cabinet, Maehara was in charge of handling the Senkaku dispute. When he was interviewed by the *Wall Street Journal* on 25 September 2010 regarding the release of the captain of the Chinese trawler Maehara explained that the Senkakau Islands had been taken by the United States at the end of the Second World War and then passed to Japan when Okinawa was returned to Japan in 1972. China had not claimed the islands until the early 1970s. The UN Convention allows fishing outside territorial waters but prohibits entry of vessels into foreign territorial waters. Maehara told the paper that when he was the transport minister in Kan's first cabinet he examined the coastguard vessel and there was evidence that the trawler had rammed the vessel. Maehara spoke of his worries about China's military spending and that figures based on data from US think tanks and the US Defense Department showed that expenditure had increased twenty times during the last nineteen years. However, according to Maehara, the budget was two-to-three times more than those figures.[9]

Maehara is considered an expert on national security policy and has a deep admiration for pre-war Prime Minister Koki Hirota. When Maehara, who is a member of the Congressional National Security Research Group and a China hawk, addressed members in the Diet in 2005 he cautioned that China's military presence was increasing and that Japan had to stand firm in order to control China's expansion.

At a press conference on 15 October 2010 regarding the Senkaku Islands dispute, Maehara took a hard line. He called China a significant threat. China's assistant foreign minister Hu Zhengyue responded by accusing Maehara of exacerbating tensions and stated that the Chinese government was not ready to arrange a meeting between Kan and Premier Wen Jiabao to discuss the issue.

Maehara entered politics first as a member of the Japan New Party, winning a seat in the Lower House as a representative from Kyoto Prefecture. He joined the New Frontier Party in 1994 and then the DPJ in 1998. Maehara became the president of the DPJ in September 2005 after defeating Kan. However, Ozawa was elected president in April 2006 after Maehara's attempt to connect the Horie Livedoor scandal to LDP politicians unraveled when the evidence proved to be flawed. Maehara served as shadow minister for security affairs and shadow minister for the Defense Agency. He was the minister of land, infrastructure, transport and tourism in Prime Minister Hatoyama's cabinet in 2009 and retained the post when Kan became prime minister.

On 7 March 2011, four days before the earthquake, Maehara resigned from his post, admitting that he had received a political donation of ¥250,000 ($3,000) from a Korean national who owned a restaurant in his hometown and whom he had known since primary school days. Although his friend had permanent residency, as a foreign national she was not allowed citizenship. Maehara was a candidate for the DPJ presidency but was defeated by Noda, who also takes a tough stand on the Senkaku Island dispute.

Maehara's debut as DPJ Policy Chief was as keynote speaker at a symposium in Washington DC where he stated that the strict regulations that limit the use of weapons by the SDF during peacekeeping operations overseas should be eased and that the lifting of the restrictions would serve to deepen US–Japan security relations. He also stated that the government would review the article in the constitution that formally bans the export of weapons.

Shintaro Ishihara: post-earthquake neo-nationalism

The detention of the captain of the Chinese trawler triggered mass demonstrations in three major Chinese cities on 16 October 2010, which quickly spread to other cities. Shouting 'Defend the Senkaku Islands!', 'Overthrow Japanese Imperialism!' and 'Fight Japan', protesters smashed Japanese retailers' shop windows. The Japanese answered with student marches in Tokyo's Aoyama Park which stands on a former military shooting range. Two thousand Japanese, waving Japanese flags, shouted 'We will not permit China to invade the Senkaku Islands! We will not permit China to invade Japan and other Asian countries!'

One of the organizers of the Tokyo rally was Toshio Tamogami who was the chief of Japan's SDF (2007–8) before he was dismissed for publishing an essay that insisted that Japan was not an aggressor in the

Second World War. He maintained that the war brought prosperity to the Asian nations which Japan had occupied and that there had been no acts of brutality. Tamogami also organized a march on 3 October in which 2,700 protesters participated.

The young neo-nationalists' mentors are high-ranking politicians as exemplified by Tokyo's governor Shintaro Ishihara whose racist and chauvinistic oratory have been well publicized by the Japanese media. Ishihara won a fourth term in office on 10 April 2011 with a landslide victory, signaling a setback in the polls for the DPJ in local elections. His win also signified a surge in ultra-conservatism among Japanese voters.

Ishihara was first known as a novelist in the 1960s, receiving Japan's most prestigious book award, the Akutagawa Prize, for *Season in the Sun*, a story about a wealthy student rebelling against post-war values by engaging in gambling and promiscuous sex. The story was adapted into a screenplay for a movie and Ishihara soon became an idol among young Japanese. When he entered as an LDP candidate for a seat in the Lower House in 1968, due to his celebrity he won a landslide victory.

In 1973, Ishihara established with other LDP politicians the 'Blue Storm Group' (*Seirankai*) which pledged unity in blood and in spirit. He served as director-general of the Environmental Agency in Takeo Fukuda's cabinet and as minister of transport in Takeshita's cabinet. Although he was a very popular politician, he was not influential enough to form his own faction. After losing the election for LDP president in 1989 he collaborated with the late Akio Morita, a founder and chairman of Sony Corporation to write the best-seller *Japan Can Say No* [*NO to ieru Nihon*] which argued that the United States regarded Japan as a subordinate and that this attitude was indicative of racism in America.

Ishihara's initial exposure to the United States was during the war when his neighborhood was strafed by American aircraft. Ishihara claimed that the aircraft flew so low that he could see the pictures of naked women and cartoon characters painted on the sides of the plane. Ishihara and writers of his generation such as Yukio Mishima[10] considered the loss of Japanese traditional values during the occupation as tragic and considered the Occupation Forces as conquering devils and Japan as a pitiful nation. Many of his remarks relate to the Japanese not having a sense of national identity.

Ishihara left politics in 1995 but returned to run for governor in 1999 as an Independent and won his first term in office. Although his controversial xenophobic, bigoted and neo-nationalistic remarks have often angered the public, his unwavering stance has cultivated among many Japanese a sense of stability during unstable times and, hence, a huge following. Ishihara's victory in April 2011 was partially attributed to his support of Japan's development of a nuclear arsenal. On 8 March 2011, just three days before the earthquake, Ishihara told *The Independent* that Japan should have nuclear weapons to offset the threat from China. He stressed that China, North Korea and Russia possessed nuclear capability and also that all of the permanent members on the UN Security Council possessed nuclear weapons. He contended that if Japan possessed nuclear weapons, China would not have disputed Japan's territorial rights to the Senkaku Islands. Ishihara claimed that Russia would also have more respect for Japan. Condemning the dismantling of Japan's military aeronautical industry by SCAP, he was in favor of lifting the ban on the manufacture and sale of weapons.[11]

Ishihara's xenophobic remarks are infamous. In 2000, addressing the SDF Ishihara stated that illegal immigrants from Third World countries committed terrible crimes in Japan and that if a disaster such as an earthquake were to occur they would instigate civil disorder. He later reconsidered his statement and apologized to ethnic Korean and Chinese residents.

The failure of Ishihara's pet project, a commercial city bank that loaned to small and medium-size businesses did put a damper on Ishihara's career as governor of Tokyo. Shinginko (New Bank) was established in April 2005 through an investment of ¥100 billion ($906 million), 84 percent of which was tax revenue from the Tokyo Metropolitan Government. Among the thirty-five corporate investors were NTT Communications, Hitachi, Orix and Kajima Construction.

Within two years Shinginko had extended loans and loan guarantees totaling ¥240 billion ($2.165 billion) despite the default of ¥2.5 billion ($22.55 million) by the end of 2005. In September 2007 Shinginko reported a loss of ¥93.6 billion ($868 million) of which ¥8.6 billion ($79.8 million) had accumulated during a six-month period. The reason given for the loss was poor assessment of collateral and mismanagement. On 20 February 2008, Shinginko applied for a government bailout of ¥40 billion ($371 million). 'Ishihara's Bank', as Shinginko was often called, received the bailout.

Ishihara's ardent admirers: Toru Hashimoto and the Osaka Restoration Group

Ishihara was instrumental in the adoption of the *New History Textbook* by the Tokyo Metropolitan Board of Education in 2004 and in ensuring that teachers in Tokyo schools who refused to urge students to sing the national anthem and salute the flag were reprimanded and punished for non-compliance. One of his biggest admirers was Osaka Prefecture governor Toru Hashimoto who won the gubernatorial election on the LDP ticket on 29 January 2008 on a platform that pledged to reduce Osaka's FY 2007 government debt of ¥5 trillion ($42.6 billion) by cutting personnel and salaries for prefectural government workers, including police officers and teachers. He also pledged during his campaign to stop the release of deficit-covering government bonds and that the prefecture's budget would not exceed revenue. However, after only six months in office, the new governor was pressured by the prefecture's bureaucracy to revise his stance on issuing bonds. In June, Hashimoto acknowledged that it was necessary to release a minimum of ¥10 billion ($85.2 million) worth of bonds to shore up the fiscal budget for FY 2008.

A week after he became the governor, on 6 February, the former lawyer and TV personality met with 500 prefectural workers to announce they were effectively working for a bankrupt company. In April, Hashimoto put native English speakers who were working in prefectural high-schools on short-term, four-month contracts, despite a letter of protest delivered to him by their union.

Hashimoto established the Osaka Restoration Group whose members were ultra-conservative and older, former members of the LDP. Hashimoto's group also expected to introduce textbooks that focus on patriotism and traditional Japanese social values.

Hashimoto fully supported the 2007 law that punishes public school teachers who do not follow orders issued by a principal to stand and sing the national anthem. Since the beginning of his term in office, the governor had been locked in an ongoing battle with teachers who opposed the law. In May 2011, an ordinance forcing teachers and other school staff to stand and sing the national anthem was submitted by the Osaka Restoration Group to the prefectural assembly. The ordinance passed because the Osaka Restoration members held fifty-seven of the 109 seats in the assembly. A separate ordinance was submitted later in the year, indicating the penalties teachers and staff would receive if they did not comply with the ordinance, which is the first to be issued by a prefectural government. Hashimoto defended the first bill when he spoke to reporters in May:

> It's not unreasonable to punish public school teachers who reject orders from a principal to stand up. If they don't like it, they can quit. It's a problem of organizational management, not a question of individual thought or consciousness.[12]

6

Pork-Barrel Patronage in the Prefectures: the Proliferation of Nuclear Power Plants

On 6 May 2011 Prime Minister Kan called for the immediate suspension of operations at the Hamaoka Nuclear Power Plant because it sits directly above the Tokai fault line where seismologists predict an earthquake over the magnitude of 8.0 could hit at any time during the next thirty years. Altogether there were five reactors at Hamaoka. No. 1 and No. 2 reactors had already been decommissioned since 2009. A sixth reactor has been under construction since December 2008.

The plant is located in Omaezaki City, Shizuoka Prefecture, and operated by Chubu Electric Power Company (CHUDEN). The utility serves customers in the Tokai region which include Aichi, Shizuoka, Gifu, Yamanashi, Fukui, Niigata and Ishikawa prefectures. Tokai is Japan's major industrial and manufacturing region with densely populated cities where economies are supported mainly by the manufacturing sector. Car manufacturers with extensive operations are Toyota Motor Corp., Honda Corp., Mitsubishi Motors and Suzuki. Yamaha, Kawasaki Instruments and Bridgestone Tire Company are also among the manufacturers that support the region's economy.

Kan's directive was spurred by the public's reaction to the ongoing nuclear crisis at Daiichi and the need to quell their growing anxieties regarding the possibility of crises occurring at other nuclear reactors located throughout Japan and to reassure them that their safety was paramount to government policies regarding the use of nuclear power. It was also a last-ditch effort to hold on to political office and it worked temporarily. The order for the shutdown invited 66.2 percent support from the public. A public opinion poll taken by Japanese dailies revealed that 47 percent of respondents also supported the reduction of the number of reactors. Kan's public approval rating, which was 17 percent before

11 March, rose to 28.1 percent. On the other hand, opposition to his cabinet stood at 57.4 percent.

However, Kan's cabinet ministers, manufacturers in the Tokai region and CHUDEN regarded the measure as entirely unrealistic because similar to the Tohoku region, the decrease in power would significantly affect operations. Furthermore, CHUDEN was in the process of considering methods to transfer supplies of electricity to the Tohoku region, which was experiencing a 25 percent decrease in power.

On 8 May, the board of directors of CHUDEN, which operates only one nuclear power plant, met with Kan to try to dissuade him from shutting down the Hamaoka plant. Their reasons were: (i) CHUDEN would be unable to provide a stable supply of electricity if nuclear power generation were curtailed, power in the Tokai region would be cut by an estimated 12 percent, and during the summer months supplies would decrease by a further 3 percent; (ii) CHUDEN would be unable to provide extra electrical power to TEPCO and would be forced to ask Kansai Electric Power Co. (KEPCO) for assistance. Since CHUDEN would no longer be able to supplement TEPCO's power supply, TEPCO would also have to procure from KEPCO the electric power that CHUDEN was going to provide before the Hamaoka suspension; and (iii) CHUDEN's costs would increase to more than $3 billion if operations at thermal power plants were to substitute for nuclear power which could cause the utility to go into debt.

On 8 May, three days after Kan made his request to CHUDEN, Deputy Chief Cabinet Secretary Yoshito Sengoku stated emphatically during an NHK talk show that the government had no intention of shutting down nuclear power plants except for Hamaoka. He reassured the audience that the government was assessing the other locations where the reactors were situated and that currently the other sites were considered low-risk areas. He also pronounced that reactors located on the Sea of Japan in particular were of low risk. Sengoku appeared to be overlooking the fact that the earthquake that hit Niigata in 2007 caused extensive damage to Kashiwazaki-Kariwa Nuclear Power Plant, which is located on the coast of the Sea of Japan near a submarine fault line.

Sengoku, as the foreign minister in Prime Minister Hatoyama's cabinet in 2010, had accompanied a delegation to Vietnam to meet with government officials and to support the establishment of a joint venture between Japanese industries and Vietnamese businesses to build nuclear power plants. The trade mission, arranged by METI and the Ministry of Foreign Affairs included executives from Hitachi, Toshiba, Mitsubishi Heavy Industries, TEPCO,[1] KEPCO[2] and CHUDEN.[3]

Banri Kaida, Kan's METI minister also reassured Makoto Yagi, KEPCO's president and the chairman of the Federation of Electric Power Companies (FEPC),[4] which promoted nuclear energy as a national energy policy, that the resumption of the operations of nuclear power plants which had been suspended for regular safety checks was primary to ensuring stable power supplies. Kaida stated that, excluding Hamaoka, the government considered that there were no safety issues at the nuclear facilities and that every effort would be made to persuade local authorities to resume operating the plant after an inspection.

On 9 May, CHUDEN agreed to suspend operations at Hamaoka's remaining reactors and on 13 May began a cold shutdown. One of the reactors contained MOX fuel.

On 10 May, Kan's Finance Minister Yoshihiko Noda warned that the power cuts by both TEPCO and CHUDEN would impact negatively, not only on manufacturers but also on employment conditions in the region. He voiced concerns about the domino-effect on Japan's macro-economy.

On 26 May 2011, Hiromasa Yonekura, the chairman of Nippon Keidanren (Japan Business Association), the most powerful business lobby group in Japan, promoted the review of nuclear safety regulations and called for the formation of an international inspection organization for the nuclear power industry. But he also criticized Kan's handling of the nuclear crisis and for pledging the halt of future construction of plants and proposing that government policy shift from nuclear to the development of alternative energy sources such as solar power. Yonekura stressed that solar power was not a reliable alternative and that nuclear power was vital to ensuring sufficient global energy supplies.

Keidanren's members include the *keiretsu*, the utility companies, energy companies, large construction firms, manufacturers and retailers. The association consistently promoted nuclear energy as one of its ten priorities. Keidanren conducts surveys of political parties to ascertain if they accommodate the business community's priorities in their policies and the degree of effort to implement these priorities. During the LDP's era fifty-four nuclear reactors had been constructed, placing the party in Keidanren's good graces.

The business lobby group was disappointed with the DPJ's lack of enthusiasm for nuclear energy until 2005 when the party began considering nuclear power generation as a temporary energy source. The DPJ was given higher marks for compliance in 2006 when Masayuki Naoshima, the DPJ director of the DPJ Policy Research Committee, replied to queries from Keidanren's vice-chairman Tsunehisa Katsunuma, who also was TEPCO's vice-president, that nuclear energy was essential to ensuring Japan's energy supply. Katsunuma was serving as TEPCO's chairman in 2011.

Naoshima later served in Prime Minister Yukio Hatoyama's cabinet as the minister of economy, trade and industry in 2009 and directed the nuclear power administration. He also attended the delegation of big businesses engaged in nuclear power to Vietnam to promote the alliance with Japanese businesses for the construction of nuclear power plants in Vietnam. In 2010, as METI minister in Kan's cabinet, he designed the 'Basic Energy Plan' which promoted the construction of fourteen more nuclear reactors during the following ten years. Masashi Fujiwara, a DPJ member in the Upper House, who was employed as an electrical engineer at KEPCO and who served as secretary-general of the company's trade union, also contended that since 2007, his party considered nuclear power energy to be fundamental to a stable energy supply.

Kan's sudden deviation from the DPJ's 'Basic Energy Plan' prompted local officials in other prefectures where nuclear reactors were located and where serious accidents had occurred to request the suspension of operations of those plants for full safety inspections by NISA. Hiroshi Aida, mayor of Kashiwazaki in Niigata Prefecture which hosts a plant that was seriously damaged by an earthquake in 2007, stated that Kan's decision was a bold one. Conversely, Hideo Kishimoto, mayor of Genkai in Saga Prefecture which hosts a plant carrying MOX fuel, regarded Kan's decision as hasty.

The temporary suspension of operations for an estimated forty-two out of the fifty-four commercial power plants during the summer would acutely affect local economies, which during Japan's post-war period have relied heavily on public works projects such as the construction of dams, roads, municipal infrastructure and nuclear power plants to support local economies and employment. Shizuoka governor Heita Kawakatsu, while praising Kan's decision, also requested that the government assume responsibility for any negative impact on the local economy from the suspension of Hamaoka. Shigeru Ishihara, the mayor of Omaezaki, also requested during a meeting with Kaida that the state promise full financial support for the 1,200 people who worked at the plant and to guarantee that the city's budget, which relied mainly on hosting Hamaoka's operations, also received subsidies.

On 24 May, Hideaki Omura, governor of Aichi Prefecture where the major automobile manufacturers have big factories, also met with Kaida to request that government ensure a stable supply of electricity. He maintained that industries in the Chubu region should not have to bear all of the additional costs associated with the suspension of Hamaoka.

Fumio Kawaguchi, the head of Chubu Economic Federation, the region's primary business lobby group, also requested that government take the responsibility for providing a stable supply of electricity.

Pushing nuclear in needy prefectures

At the end of the war Japan's primary industry was agriculture. In the late 1950s when Japan began its period of rapid economic growth, thousands of young male and female high-school graduates living in farming and fishing villages left for the Tokyo–Osaka–Kobe regions to seek better-paying jobs in retail shops, factories and small businesses. The mass migration resulted in the loss of population in communities outside of the industrialized areas and the overpopulation of big metropolises. Kakuei Tanaka's 'Plan for the Remodeling of the Japanese Archipelago', while addressing the problem, also set the stage for government policies such as the Land Development Policy that called for investment (FILP) in public works throughout Japan to revitalize the less populated rural areas. Tanaka's book also promoted the return of the workers who had emigrated from rural areas to their hometowns for a healthier, pollution-free way of life and to start new businesses.

Since the impact of the first oil crisis in 1973 which occurred during Prime Minister Tanaka's administration, the development of nuclear energy has been a national priority. In order to continue economic expansion and overtake the US economy the Japanese government had to reduce industries' reliance on fossil fuels, 80 percent of which were imported. In 1974, the 'Three Power Source Development Law' was enacted to subsidize local communities that were willing to host nuclear power plants. The law called for a tax, which was included in consumers' utility bills, to be used by government to subsidize public works in communities that hosted the plants.

Since 1966 local governments hosting nuclear reactors have received altogether ¥2.5 trillion in nuclear-related subsidies. ¥915 billion in subsidies was distributed after the law was passed and the host cities, towns and villages charged the plant operators ¥892 billion in fixed property taxes. The utility companies are said to have donated ¥53 billion to the local governments. However, according to the *Mainichi Shimbun*, which conducted the survey in August 2011 supported with ANRE data, the total amount could well be higher because local governments were reticent to release data that confirmed the total sum of nuclear-related fixed property taxes and donations. Hokkaido and the twelve other prefectures where nuclear power plants are located have received approximately ¥675 billion in fuel tax revenue from the utilities.

The state energy agencies, namely MITI and MITI's ANRE and the ministry's affiliated institutions, had been engaged since the 1950s in the development of atomic power but after the oil shock MITI, together with the regional utility companies and the heavy industries involved in nuclear power such as Mitsubishi, Toshiba and Hitachi, aggressively chose sites for the construction of nuclear power plants in rural, depopulated municipalities with inadequate tax bases, where the civil societies were considered weak and where the need for subsidies to support local small businesses and maintain employment would promote the acceptance by local authorities of nuclear power plants in their areas. Japan's arable land is only 15 percent of the total and agriculture, which is highly subsidized and protected by government, accounts for only 1.3 percent of the GNP.

Despite the efforts of anti-nuclear activists who tried to structure policies to incorporate references to Hiroshima and Nagasaki, from the 1970s onward the majority of Japanese were pro-nuclear, especially in rural areas from where the young workforce had departed to search for a higher standard of living in the industrialized areas and where the population was shrinking. The local authorities came to accept central government directives usually with little protest because of the ministerial controls over local government officials and because, if the directives by the state agencies were refused, the communities might forfeit future subsidies for public works projects on which they had come to rely. The communities became addicted to the subsidies as the local governments connected hosting power plants with the subsidies for the future construction of schools, free and improved welfare services and lower local taxes. It was not until the late 1990s after a series of accidents occurred at power plants that the Japanese began to recognize the implications of building nuclear plants in earthquake-prone zones and to question the safety of the installations.

On 15 February 2011 the Bank of Japan upgraded its general assessment of the macro-economy, indicating that the country was gradually emerging from recession. However, the 'emergence' relied heavily on the increased demand for exports in emerging markets from the large exporters and not on domestic consumption, which accounts for 60 percent of the GDP. In an attempt to raise the consumption of durable goods, the government supported purchase of eco-friendly white goods, home electronics and automobiles through tax incentives.

Government investment in all regions had decreased from the previous year. Capital investment had improved in some regions, albeit at very low levels. While retail sales were flat for most regions the investment in homes and real estate, which had been deteriorating, was increasing slightly in some regions. Employment conditions in all of the regions remained severe.

The BOJ economic reports for the nine regions released in January 2011 gave mixed assessments, which were similar to the assessments released in October 2010:

1. **Hokkaido**: In October, the economy was said to be continuing to improve, although conditions as a whole remained 'severe'. In January, the economy 'seems to have paused recently, although it had continued to pick up'.
2. **Tohoku**: In October, the economy was 'picking up'. In January, the economy was picking up in general, 'although the recovery seems to be pausing particularly in manufacturing'.
3. **Hokuriku**: In October, the economy was continuing to pick up, 'although there remains some severity in economic conditions'. In January, the economy was recovering moderately but 'improvements have slowed'.
4. **Kanto-Koshinetsu**: In October, the economy was continuing to pick up as a whole, but 'improvements have slowed. Differences among regions and industries also remain.' In January, the BOJ reported that the economy was recovering moderately but that 'the recovery seems to be pausing' and that 'differences among regions and industries also remain'.
5. **Tokai**: In October, the economy was picking up but 'the pace of improvement seems to have recently slowed sharply'. In January, 'The economy has paused.'
6. **Kinki**: In October, the economy 'with some severity in employment' was recovering moderately. In January, 'The economy has recently paused.'
7. **Chugoku**: In October, the economy was recovering moderately but 'the pace of recovery has been slowing'. In January, 'The recovery in the economy seems to be pausing.'
8. **Kyushu-Okinawa**: In October, the economy was recovering moderately but employment and wage conditions remained 'severe'. In January, the economy continued to recover moderately but the decline in demand from foreign markets was predicted because of increasing inventories.
9. **Shikoku**: In October, the economy was picking up moderately 'although economic conditions as a whole remain severe'. In January, the BOJ reported that the pickup 'seems to be pausing'.

The regional assessments are general overviews. The prefectures within these regions with agriculture and fisheries as primary industries and which are less populated relied more on public works projects to support local economies. A good number play host to nuclear power plants.

Fukushima, Aomori and Fukui are examples of prefectures which host plants and are the recipients of government subsidies.

Fukushima Prefecture

This prefecture is the fourth-largest farmland area in Japan and primary industries are large-scale agriculture and fisheries. Fukushima is one of the top producers of rice, peaches, apples, pears and cucumbers. The secondary industries with factories located near the capital Fukushima City include information and telecommunication systems, electronic parts, electrical machinery and chemicals. Fukushima's population in 2005 was 209 million but was in decline due to infertility nationwide and the emigration of jobseekers.

Fukushima's economic conditions prior to the earthquake were essentially unchanged from the previous year. In 2010, retail sales in department stores remained flat due to consumers tightening purse strings and postponing purchases. Supermarket sales continued to fall. The purchases of eco-friendly white goods, electronics and automobiles had increased in 2010 because of government subsidies and the change to digital in the region. Nevertheless, during the first two months of 2011 sales of automobiles decreased over 30 percent due to the termination of the tax-break scheme.

Residents continued to economize on travel and leisure activities because of severe wage and employment conditions. The rate of the deterioration in capital investment in installations and equipment was slowing from 2009 because of the refurbishment of old facilities. However, the sentiment among small and medium-size business owners was pessimistic due to the appreciation of the yen, the instability of the US economy, deflation and Japan's general economic stagnation. Personal savings continued to increase as concerns mounted about future economic conditions. Investment in housing and real estate had deteriorated from 2009.

Former Fukushima governor Eisaku Sato announced in February 2001 that he would never allow mixed plutonium and uranium oxide fuel (MOX) to be used in the Daiichi Power Plant. He and local residents managed through petitions and public referendums to delay the addition of MOX to the fuel rods until 2006 when Sato resigned after eighteen years in office because of the criminal charges that he had accepted bribes from a subcontracting firm. He placed the blame squarely on METI and NISA for the 2011 Daiichi disaster.

MOX is mixed oxide of plutonium and uranium. Plutonium is separated first from spent nuclear fuel at reprocessing plants in Japan and in France. It is then mixed with uranium and concocted into fuel for light water reactors. The advantage of using MOX is that it is a less expensive fuel because the waste can be reprocessed and reused. On a less positive note, plutonium is more combustible than uranium and more difficult to control than uranium fuel if accidents occur. MOX is highly toxic and can stay in the environment for thousands of years. A reactor with a containment vessel loaded with 100 percent MOX is more likely to fail when there is a core meltdown. Although the pluthermal program was initiated in 1997, MOX fuel was not commercially used until 2009 because of ongoing problems regarding adapting older reactors to carry the fuel rods.

For better or for worse, Fukushima's two nuclear power plants Daiichi and Daini provided hundreds of jobs and encouraged large government investment in the prefecture. On 16 February 2010, Yuhei Sato, who succeeded Eisaku as governor, stated that he would accept pluthermal power-generation (MOX) in No 3. reactor if Daiichi was properly protected against earthquakes and aging. TEPCO began to load MOX, which was stored at Daiichi, into No. 3 reactor on 27 August. When the earthquake hit, MOX accounted for 6 percent of the fuel in the fuel rods.

Three months after the earthquake, in June, *Asahi Shimbun* and researchers specializing in local government from Fukushima University conducted a survey of the 407 evacuees. In response to questions regarding the safety of nuclear power plants, 66 percent considered that nuclear power was safe prior to the earthquake. However, following the earthquake 70 percent of the subjects were opposed to nuclear power. Forty-six percent felt that the accident could have been prevented and that a major cause was that the reactors were old. Surprisingly, 52 percent had been in nuclear reactor-related jobs and 75 percent considered that nuclear power plants had helped the local community's economy.[5]

Aomori Prefecture

Aomori is located in the Tohoku region and is the northern-most prefecture on the main island of Honshu. With a population of about 1,437,000 people it is the twenty-eighth largest prefecture in Japan. In terms of land mass it is the eighth largest prefecture. The primary industries are forestry, aquaculture and agriculture. Aomori ranks first among the forty-seven prefectures for apples, yams, burdock root and garlic, squid, pond smelt and whitebait.

In early 2011, employment and wage conditions were severe with about 0.32 contract jobs available per applicant. Prior to the earthquake, consumption continued to be weak as residents were saving income rather than spending it. Retail sales at department stores and small retail outlets were depressed and sales of food and beverages were also weak. Similar to Fukushima, sales of eco-friendly automobiles, electronics and white goods decreased from a year earlier because of the withdrawal of tax-breaks and subsidies that had served to support sales.

Although the amount of bonds issued by central government for public works projects had decreased, overall, due to the bonds issued by the local authorities, the subsidies were higher than the previous year.

Corporations continued to ration capital spending due to the instability of the rate of demand for exports from emerging markets. The production of electrical equipment and iron and steel increased but production for all other industries was slow.

Personal savings increased while corporate savings decreased. There was a sharp decrease in personal and corporate loans but the loans to the public sector increased. The average contract interest rates, short and long-term, fell from the previous year.

Aomori's no. 1 industry: nuclear power

Aomori, which had been considered for oil storage facilities until the oil shock in 1973, currently hosts one nuclear power plant with two reactors. However, with four more power plants currently under construction, the prefecture will be host to more nuclear-related installations than any other prefecture in the country.

The nuclear power industry has become a dominant part of Aomori's economy and nuclear facilities are located in some of the most scenic areas along the Pacific coast. The integration began in July 1984, when FEPC asked Aomori Prefecture and Rokkasho-mura for permission to locate nuclear facilities in the village. In the 1980s, as the nuclear energy programs expanded along with the Japanese economy, the government designated the area for a uranium enrichment plant and for a low-level radioactive waste disposal facility. In 1980, the electric power companies established Japan Nuclear Fuel Ltd. (JNFL) to commercially recycle spent fuel. In April 1985, a basic agreement on the location of the nuclear facilities was reached. JNFL financed and operated the Rokkasho Uranium Enrichment Plant which began operations in 1992. The technology was developed by the Power Reactor and Nuclear Fuel Development Co., which was renamed the Japan Nuclear Cycle Research Institute (JNC) when it was reorganized as a Special Corporation by MITI in 1998.

JNFL's[6] second installation, the high-level Radioactive Waste Storage Center, was completed in 1995 and began operating the same year. The storage facility receives spent fuel from nuclear power plants located throughout Japan. However, its safety practices regarding storage of the spent uranium fuel came into question when in September 1999 radioactive liquid leaks were discovered on waste drums shipped from Fukushima Daiichi and Daini and Hamaoka nuclear power plants.

In 1998, JNFL received a request from FEPC to begin studies regarding the development of a MOX fabrication plant (J-MOX). Subsequently, in 2001, the Aomori government received a request from METI to build Japan's first nuclear fuel reprocessing plant in Rokkasho Village.[7] The complex, to be financed and operated by JNFL, includes a MOX fabrication plant, a uranium enrichment plant and a high-level nuclear waste facility. The investment in the development of the complex had already ballooned to $20 billion by 2001.

In 2004, the news that METI had failed to release a 1994 report revealing that the cost of recycling nuclear waste was more expensive than burying it made Aomori officials doubt the viability of the reprocessing plant. On 8 July 2004, Aomori governor Shingo Mimura lodged a protest with ANRE regarding the concealment of the information. ANRE director-general Nobuyori Kodaira[8] visited Aomori to apologize personally for the mishandling of the data a few days following METI's admission of failing to make the information public. In March, Kazumasa Kusaka,[9] ANRE's previous director-general, had contended at a closed-door session before a government panel of experts in the Diet's Upper House that no such data existed. According to Kusaka, the only costs that were presented were estimates for the reprocessing of the waste. The panel decided not to release the data to the public. Following his term in ANRE, Kusaka was appointed MITI vice-minister of international affairs.

In 2002, taking into consideration the costs of construction and equipment, ANRE had estimated that the cost for reprocessing the fuel in-house was 1.5 times more expensive than importing it from Great Britain or France, or ¥350 million ($2.8 million) per metric ton of spent fuel, and twenty times the cost of burning uranium.

METI's vice-minister Hideji Sugiyama had also apologized at a news conference on 5 July, stressing that METI had not known of the data and that the ministry never had any intention of concealing it.[10] Following METI's apologies, representatives from FEPC met with Aomori government officials as well to explain why in 1996 it had concealed information that indicated that the cost of burying spent fuel was about 30 percent less than recycling it. FEPC claimed that the data were not considered important enough to publish and that there was no need to review the national nuclear fuel recycling policy or to change the current strategy for recycling spent fuel.

Ultimately, in April 2005, Aomori governor Shingo Mimura acquiesced and announced that the prefecture had approved the construction. When the governor signed the agreement together with the mayor of Rokkasho-mura and the head of JNFL he urged FEPC to speed up its efforts to implement the MOX program.

On 6 November 2006, JNFL announced that it had successfully produced its first uranium–plutonium mixed oxides. The JNFL facility will reprocess plutonium and residual unburned uranium that is extracted from the spent fuel to be fabricated into MOX fuel, a technology that was developed by the Japan Atomic Energy Research Institute (JAERI). On 20 April 2010, five years after JNFL had submitted an application, it received the go-ahead. On 13 May 2010, JNFL received the license from METI to build and operate the 130 metric tons per year J-MOX plant. Mitsubishi Heavy Industries was the builder.

Despite the growing public mistrust of government information about the safety of nuclear power plants triggered by TEPCO's admission in 2002 that it had falsified safety reports, the construction of the plant was underway in October 2010 after a three-year delay due to a reassessment of seismic criteria. The cost was estimated to be ¥190 billion ($27.33 billion). Again, due to safety issues and protests lodged by antinuclear activists, commercial operations were delayed. It was anticipated that the plant would be operating commercially by 2012. Meanwhile, tons of spent fuel were accumulating in nuclear reactors around the country waiting to be reprocessed.

JNFL's main office is located in Rokkasho-mura. The ten domestic electric power companies and seventy-seven other firms are the stockholders. TEPCO, the largest shareholder, promotes pluthermal reactors and MOX fuel on its website as a means of ensuring a reliable supply of energy and states that 'electric power companies are making various efforts to obtain broad public acceptance of this new power generation method'.[11] The website also informs the public that the company had loaded MOX into Daiichi's No. 3 reactor.

In 2000, the Radioactive Waste Final Disposal Act was passed and METI established the Nuclear High-Level Waste Management Organization (NUMO) in the same year for choosing deep geological storage sites for the spent fuel and for constructing and operating the facilities. Although from 2002 applications for the selected sites were distributed to thousands of towns none of the locations responded. On its corporate website TEPCO also promotes NUMO's search for sites that are deemed low risk in terms of volcano eruptions and earthquakes.[12]

In May 2010, JNFL also received a permit to construct a temporary storage facility for spent fuel for TEPCO's company the Recyclable-fuel Storage Company (RFS) in Mutsu where fuel from Japan Atomic Power Company (JAPCO) nuclear power plants will be reprocessed at the JNFL plant. JNFL also operates long-term storage facilities.

METI's JNC and MEXT's JAERI were consolidated in 2005 into one organization, the Japan Atomic Energy Agency (JAEA). The agency operates a research and development institute in Rokkasho as well as several other nuclear facilities. NISA has an office for nuclear safety inspectors.

The Higashidori Nuclear Power Plant, which began operating in 2005, is located in the small town of Higashidori. Tohoku Electric Company (TODEN) operates the plant. TEPCO received a permit from METI on 30 December 2010 to begin construction on two more reactors in the Higashidori complex. METI vice-minister Motohisa Ikeda personally handed to TEPCO president Masataka Shimizu the permit which read: 'We hope you will take all possible measures to secure safety as Japan is an earthquake-prone archipelago.'[13] TEPCO began construction in January 2011. The advanced boiling water reactors were expected to be commissioned by March 2017.

Higashidori Village's economy was primarily commercial fishing before TEPCO arrived in 1992 to ask the local government for permission to build a nuclear reactor. Since then, TEPCO has injective massive funds into the area and the nuclear reactors have become the town's primary industry. During the past decade, the Shimokita area where Higashidori is located has seen the emigration of hundreds of construction workers who have been laid off by contractors. The construction of the other reactors at Higashidori power plant complex brought subsidies for the construction of a new elderly care home, a new clinic, a school and a childcare facility. However, the population has dwindled to around 7,300 and the sprawling school sports facilities paid for by subsidies are used by only about 600 students in the primary and secondary schools.

In 2010, 46 percent of Higashidori's $94 million budget came from government subsidies and the economy would suffer significantly if government subsidies dropped; the construction of new plants to replace the older ones would bring additional subsidies and encourage employment in Higashidori. The local authorities felt very positive about the sustainability of their economy. After the 11 March earthquake, the work on Tohoku Electric's new reactor was halted temporarily but operations were planned to resume after NISA inspectors were satisfied that a thorough inspection had been performed. The town's Nuclear Power Countermeasures Division met to consider other options than nuclear but the consensus was that Higashidori would be unable to discontinue the plants' operations because of TEPCO's massive investment in the area. In other words, local residents were hoping that the construction of the three remaining plants would continue.

Oma is a village near Higashidori where the economy was also constrained and the 6,300 residents wanted to receive the same treatment as Higashidori. In 2008, METI gave a permit to J-POWER (the former EPDC) to operate a plant in Oma. Currently under construction, the pluthermal reactor with 100 percent MOX fuel is expected to be in commercial operation by 2014. The subsidies provided a school, a care home for the elderly and a fishing facility. In May 2011, the majority of residents declared that the Oma plant was an absolute necessity and hoped that construction would continue. There was also the hope among the residents that more reactors would be constructed in the future.

Local government officials claim that central government view local authorities as 'beggars' and that they must court ministerial favors by offering gifts to top government officials They also complain that they are 'slaves' to central government. Governor Mimura, who won a third four-year term in office on 4 June 2011, campaigned for safe nuclear energy and promised to put in place an independent panel of experts. However, his campaign avoided mentioning the power plants that were currently under construction. Running as an Independent, he was supported by the LDP and the New Komeito Party. Mimura's opponent in the election, Takashi Yamauchi, a former prefectural assembly member and also an Independent, was supported by the DPJ. Yamauchi promised voters that he would freeze the existing plans for the construction of more nuclear reactors in Aomori. He also called for the end of nuclear powered energy.

Pork-barrel patronage is a primary characteristic of the Japanese political economy and the elected officials who represent poorer regions can remain in office for many years on the condition that they bring public works contracts to local businesses.

The princes of pork-barrel patronage: Kakuei Tanaka and Noboru Takeshita

Tanaka and Takeshita are credited with setting the standards for pork-barrel patronage in post-war Japan. Their interpersonal relationships with the construction industry, MOC, MITI and MOT facilitated a multitude of land deals and public works infrastructure projects, reaping enormous profit for both big and small businesses and large sums for their families and for the LDP treasure chest. As significantly, due to the loyalty of their constituents in their remote, agrarian-based prefectures both politicians remained in political office well after they were implicated or indicted in corruption scandals.

The godfather

In the early twentieth century, Niigata was a densely populated, agricultural prefecture which supplied the Japanese with the finest rice and sake. But, as the country industrialized the younger generations left for the big cities and gradually, the population decreased. At the end of the Second World War, Niigata ranked among the poorest prefectures in Japan. Less than a decade later, Tanaka had become a major figure in Japan's government and his generosity to LDP politician's ministry officials helped to secure for his prefecture bountiful subsidies from central government. In 1962, Niigata ranked fifth among the forty-seven prefectures, receiving ¥12.1 billion ($33.6 million) from the national budget. In 1965, Niigata ranked fourth with ¥24.1 billion ($66.94 million) and in 1970 Niigata had achieved third position with an allocation of ¥53.3 billion ($177.7 million). By 1983, Niigata ranked first in the nation for per capita expenditure for public works.[14]

Tanaka was the godfather of Japan's pork-barrel political society. His career was equally divided between policy-making in government and pork-barrel transactions. His survival record in maintaining his elite position in government despite allegations and indictments for illegal business dealings is the ultimate master class in the methods of survival in Japanese politics and in the uninhibited use of the Japan Inc. model.

Often referred to as 'the human bulldozer' (the JH was known as 'Japan's General Contractor'), besides the Tanaka Civil Engineering and Construction Company and the Nagaoka Railroad Company, Tanaka also served as the president of his synthetic fiber company Riken Chemical Industry Co. Ltd. He was the chairman and later vice-president of Echigo Kotsu, a transport company, Nihon Denken Co. Ltd., and Shin Nihon

Denken Co. Ltd. Tanaka saw to the needs of his constituents and the members of his faction during the 1960s and 1970s while accumulating personal wealth.

Tanaka's 'Plan for the Remodeling of the Japanese Archipelago' can be considered as a grand design for the nation's industrialization and decentralization. It can also be viewed as a daring way to raise land prices and to use public funds for pork-barrel patronage in the prefectures, specifically in Niigata. During the period 1967–70, three of the six ministers of construction to serve in prime ministers' cabinets were members loyal to Tanaka. From 1970–6, the seven ministers of construction and four ministers of posts and telecommunications were in Tanaka's domain. When Tanaka was the MITI minister in Sato's 1971 cabinet he forged a strong bond with MITI when he participated with MITI officials in the textile trade dispute with the United States trade representative (USTR) that culminated with Japanese textile producers receiving large compensation for the caps that had been placed on their exports to the United States. There is speculation that Tanaka's hard line with the USTR related more to his personal business interests and Riken Chemical Co. than to the defense of Japan's textile industry.

By 1980, Niigata was the beneficiary of large public works projects, courtesy of Tanaka's interpersonal networks with ministry officials, the construction, utilities, telecommunications and heavy industries. When members of the Etsuzankai petitioned Tanaka with requests for business he answered by bringing in public works projects. The funds were funneled through the association. The projects included:

1. In 1971, Tanaka initiated the construction of the Joetsu Shinkansen, Japan's third-oldest bullet train constructed at costs estimated to be $6.3 billion. The ostensible objective was to connect Niigata with Tokyo in order to promote regional development. One-third of the line travels through thirty tunnels and is said to be the most expensive railway line in Japan. From its inauguration in November 1982 the line has operated at a loss. Also, although land was purchased along local railroad lines, construction never commenced.
2. Tadami river hydro-electric power project.
3. New Shimizu tunnel.
4. A new railroad station for Nagaoka, one of the main stations on the bullet train's line.
5. Kan'estsu and Hokuriku expressways.
6. Shinano riverbed reclamation project.

7. Roads and tunnels through Niigata's mountainous terrain.
8. Kashiwazaki-Kariwa Nuclear Power Plant, located next to Tanaka's hometown of Nishiyama.
9. Electrification of the Nagaoka railroad.
10. Formation of the waste land along the Uno river.
11. Resuscitation of the Uonuma railroad.

Tanaka understood that the key to political power was making money and gifting money. The accumulation of personal wealth was integral to financing the LDP machine. In 1974, the Japanese press accused Tanaka of giving ¥3 to ¥5 million yen to all members in the National Diet and to many bureau chiefs in the ministries.[15] Tanaka's interpersonal network with construction and real estate firms were mutually beneficial and earned the companies large profits.

Tanaka established dummy companies with no staff or offices to purchase real estate in Niigata which was targeted to be the sites of the nuclear power plant, the bullet train passage, a university and land reclamation, to name a few. The profits from the land-flipping deals were extremely lucrative. There were numerous allegations against Tanaka by public prosecutors regarding kickbacks to Etsuzankai for the Tadami river hydro-electric project and for the reclamation of the Shinano riverbed, which Tanaka purchased from farmers in the 1950s. Usually he emerged unscathed from the series of allegations.

The Kashiwazaki-Kariwa Nuclear Power Plant: the art of pork-barrel patronage

Kashiwazaki-Kariwa is a classic example of land-flipping and pork-barrel patronage in the rural and poorer prefectures and why, despite serious accidents, the plant continued to operate.

Kashiwazaki-Kariwa Nuclear Power Plant (KK), which is operated by TEPCO, was constructed on land which had been purchased from Hokuetsu Paper Company in 1966 by the mayor of Kariwa, a village adjacent to Kashiwazaki, which is a major city in Niigata and in Tanaka's third district constituency. The mayor, who later confessed that he was acting on Tanaka's behalf, sold the land to Muromachi Sangyo, one of Tanaka's dummy companies which, in turn, sold it to TEPCO for far more than the original price. Tanaka's Etsuzankai realized a profit estimated to be ¥400 million ($11 million). In 1969, Tanaka convinced the government to plan the construction of a nuclear power plant on the site. FILP supplied the funds and the ministries' Special Corporations distributed the contracts and licenses to companies that were mainly in MITI's and MOC's administrative jurisdiction.

The power plant is the largest in the world and built on 4.2 km^2 of land on the coast of the Sea of Japan near the epicenter of the strongest earthquake to occur at a nuclear power plant prior to Fukushima Daiichi. When the Niigata-Chuetsu-Oki earthquake struck in July 2007, killing eleven people and injuring over a thousand, all of the reactors were shut down for twenty-one months until the damage that the plant had sustained was repaired. Two reactors went on line again in 2009.

Kashiwazaki had a population of 93,000 and the local residents were worried before the first reactor was constructed that the ground of the site was too soft for a nuclear power plant. In the KK design TEPCO only took into consideration land faults and not the faults under the ocean. The 2007 earthquake was caused by the submarine fault. Since there were no seismic guidelines when KK No. 1 reactor was approved in 1977, a government panel of experts was later assembled and they concluded that the seismic design for the plant would withstand an earthquake of the magnitude of 6.9 from the Kinomiya fault located 20 km from the site. However, some experts disputed the findings because of the other faults running close to Kinomiya and there were concerns that, together with Kinomiya, a single fault line would emerge. Nevertheless, TEPCO preferred to ignore the warning and KK reactor No. 1 was approved to withstand an earthquake of 6.9.

In 1981, the first 'Regulatory Guide for Reviewing Seismic Design of Nuclear Power Reactor Facilities' was published. After the 1995 Kobe earthquake on 17 January 1995 a new set of national guidelines was published in 2007. The new guidelines called for checks of all of Japan's nuclear power plants to ascertain whether the plants' designs would withstand the ground motion of an earthquake.

Professor Ishibashi[16] was a member of the 2006 government subcommittee which revised the existing guidelines. His proposal that the revision include the survey of active faults was rejected. Ishibashi resigned from the committee claiming that the review was rigged to favor the interests of FEPC, which had eleven of the nineteen members on the subcommittee.

All of the electric power companies submitted reports to NISA which increased the level of the magnitude of earthquakes as well as the force of the ground motion their plants would be able to endure. The companies assured NISA that they were reinforcing their plants. However, the reports were submitted too late for KK. The earthquake registered 6.8 but the ground motion was much stronger than the plant could withstand. When the earthquake struck, pipes in KK's reactors burst and containers of radioactive waste material fell, the monitoring equipment stopped working and 1,200 liters of contaminated water leaked into the ocean. A fire broke out and continued burning for several hours. The International Atomic Energy Agency (IAEA)[17] had cautioned KK's plant managers two years before the earthquake occurred that the fire-prevention measures were insufficient. TEPCO initially was willing to admit that only a small amount of radioactive water had leaked from No. 1 reactor. But on 30 July TEPCO conceded that the amount of water was closer to 1,200 liters. Tsunehisa Katsunuma, who was TEPCO's president at the time, visited KK to look at the damage. He pronounced the site in complete disarray.

KK's seven reactors became commercially operational between 1985 and 1997. The first three reactors were designed by Toshiba. Toshiba also supplied the reactors and the steam generators. Kajima constructed the reactors. Hitachi designed No 4. reactor and supplied the reactor and the steam generator. Shimizu Takenaka Co. constructed the reactor. Hitachi also designed No. 5 reactor and supplied the steam generator. Hazuma Gumi Co. supplied the reactor. Shimizu Takenaka constructed this reactor as well. Toshiba designed No. 6 reactor and together with General Electric Co. supplied the reactor. GE and Hitachi supplied the steam generator. Hazuma Gumi Co., Kajima and Kumagai Gumi were the contractors.

It is interesting to note that Kajima (founded in 1939), Shimizu (1804), Takenaka (1910) and Kumagai Gumi (1902) are elite members of the big six Japanese construction companies. Hazuma Gumi (1889) is regarded a second-tier company but with large operations as well. Traditionally, retired officials from MOC served as presidents and as executive directors in these companies.

KK accounts for about 10 percent of TEPCO's output and almost 20 percent of total generated power. It is indispensable for the provision of a stable supply of electricity in the Tohoku region. At ¥649 billion ($5.53 billion) the impact of the damage from the earthquake and shutdown of KK was enormous. The restoration costs were estimated at ¥64 billion ($545.3 million), the fuel expense alone was ¥585 billion ($4.98 billion) and the increase of the cost of purchasing supplemental power sources was ¥64 billion. There were additional losses related to the earthquake disaster and to the restarting of thermal power plants to supplement power needs.

The town that can't say 'no!'

Since 1974, directly after the first 'oil shock' the government has proclaimed that nuclear energy is a state policy and that the Japanese should fully cooperate. However, local anti-nuclear activists in Kashiwazaki-Kariwa have been protesting against the use of nuclear power and MOX and against the proliferation of nuclear weapons. There were demonstrations, petitions and referendums in Niigata before 1985 when the first reactor was scheduled to go online. In 2001, 53 percent of the Niigata electorate voted against the use of MOX fuel. Nevertheless, METI refused to budge on its long-term objectives for MOX and began looking for other prefectures (e.g. Aomori, Ehime, Saga) to build pluthermal plants. It envisioned MOX in sixteen to eighteen reactors by 2010, a vision that did not materialize because of the ongoing problems in the reactors and the resistance of local residents.

The struggle to decommission KK lost momentum, the reason relating to Niigata's need for government subsidization of the local economy and the pressure from the ministries, from TEPCO shareholders and from manufacturers with big investments in the Tohoku region on local government officials to continue to host KK.

Directly after the earthquake occurred, the mayor of Kashiwazaki issued an emergency order to shut the plant down. He reassured residents that he had taken a strong stand against TEPCO and that the plant would be safe after the repairs had been made. However, nobody was convinced because in early 2007 TEPCO admitted that it had submitted inspection reports regarding Kariwazaki and other plants it was operating with data that had been falsified. Indeed, Niigata governor Hirohiko Izumida was extremely concerned about TEPCO's falsified reports dating back to the 1980s and on 15 February 2007 he personally visited the office of Akira Amari, Minister of Economy, Trade and Industry (2006–8), to present a letter requesting that NISA ensure the safety and security of KK. Amari promised the governor that NISA would issue orders to all of the electric power companies to re-evaluate the safety of their nuclear power facilities. NISA issued a special order to TEPCO.

The safety regulation guidelines are in the jurisdiction of the NSC, an independent regulatory body established in the Cabinet Office to oversee NISA. However, in an article for the *International Herald Tribune/Asahi Shimbun* on 11 August 2007 Professor Ishibashi, who had warned that a major earthquake would strike near Kashiwara-Kashiwa, questioned the NSC's independence from government intervention since NISA officers sit on the panel.[18]

In 2008, TEPCO performed an inspection of the plant which included the buildings, equipment and other facilities and submitted the data to NISA. In October 2008, NISA reported to the NSC its assessment of TEPCO's report on Kashiwazaki Kariwa, stating that it was 'appropriate'.

In 2009, the NSC approved NISA's recommendation to restart No. 6 and No. 7 reactors.

Despite the protests from anti-nuclear residents, the majority of Kashiwazaki City and Kariwa Village residents felt obliged to accept a nuclear power station that sat on soft ground near a long fault line because the government had generously poured more than $2 billion in subsidies into the local economy and the plant employed hundreds of people. Ultimately, the residents and local authorities were controlled by the concerns that if they refused to accommodate KK, government support for other public works would cease.

In August 2011, a government panel assigned to investigate the causes of the Daiichi accident reported that the interviews conducted with sixty officials from NISA, the Cabinet Secretariat, the NSC and several hundred other officials confirmed that NISA had discouraged the Niigata government from holding twin-disaster drills in November 2010 that would prepare residents for a simultaneous occurrence of an earthquake and nuclear event. NISA advised against the drill because it would trigger fear among the local population. Consequently, the Niigata government staged a snow-nuclear disaster drill instead.

Noboru Takeshita: bringing home the bacon

Prime Minister Noboru Takeshita (6 November 1987–3 June 1989) was Tanaka's key fund-raiser while he was in Tanaka's faction. He tirelessly combed the countryside collecting money for the LDP treasury. While he served in cabinet posts Shimane Prefecture received numerous public works projects. Takeshita was chief cabinet secretary from 1971–4 in both Eisaku Sato's and Kakuei Tanaka's cabinets. When he was minister of construction in Takeo Fukuda's 1976 cabinet government subsidies to his prefecture soared. Under Tanaka's tutelage, Takeshita forged tight networks with the construction industry and the Ministry of Construction. The large general contractors were particularly fond of Takeshita because his projects brought lucrative contracts.

Takeshita was the minister of finance in Masayoshi Ohira's 1979 cabinet and in Yasujiro Nakasone's cabinet from 1982–6. Takeshita was a key participant in the Plaza Accord in 1985 which triggered the yen's appreciation against the dollar and other foreign currencies and the beginning of Japan's asset inflated 'bubble economy'.

Shimane Prefecture is the second least populated prefecture in Japan. The land mass is 6.708 km^2 but the population is a meager 0.74 million. The prefecture, which is located in the southwest of Honshu, shares borders with Tottori, Yamaguchi and Hiroshima prefectures. Its coast lies on the Sea of Japan. The terrain is mostly mountainous with only 6 percent of arable land available for agriculture, of which 80 percent is devoted to rice farming. The primary industry is fisheries. Secondary industries are machinery, iron, steel and electronic devices. Shimane also includes the Oki Islands and it claims territorial rights to the South Korean-controlled island of Takeshima.

In 2010, the BOJ's general assessment of Shimane's economy showed the same lopsided conditions that had been plaguing Japan's macroeconomy since the 1990s. On the one hand, there was a slight improvement in capital investment in equipment and installations but, on the other hand, employment and wage conditions remained severe. Similar to the conditions in Fukushima and Aomori, the consumption of durable goods was supported by government tax incentives. Business sentiment was improving because of the increase in the exports of manufactured goods to China. Government subsidies for public works projects had decreased overall. Investment in new home builds and lots had continued to decrease. The average contracted long-term interest rate offered by banks in Shimane was 2.2 percent and slipping slightly, which replicated the interest rates in the other regions.

Takeshita's commitment to bringing home as many public works projects as possible was logical because the projects served to keep his small business constituents in Shimane and the construction industry satisfied and the LDP coffers full. Takeshita was able to retain his seat in the Diet well after he resigned because of his continuous effort to supply his constituents with subsidies, which funded about 60–70 percent of public works projects. During the 1970s and 1980s Shimane's infrastructure was substantially upgraded to include three airports, tunnels and four expressways, land reclamation projects and numerous dams. There is one national university and one prefectural university. It is questionable whether the funds spent during Takeshita's years justified the extent of the construction because of the intermittent use of some of the infrastructure. The airports and railway stations are virtually empty of passengers except during peak tourist seasons and the care and refurbishment of these and the other facilities weigh heavily on local tax revenue.

By the mid-1990s Shimane citizens began to protest against the construction of new dams and land reclamation projects that had been planned during Takeshita's era such as the Lake Nakumi reclamation project that would have created farmland but would also have destroyed fishing grounds. However, with the termination of projects came new appeals from residents for the substitution of other public works projects to revitalize the area so that government subsidies would continue unabated.

Shimane Nuclear Power Plant: public works perpetuate more public works

Takeshita's pork-barrel patronage brought Shimane a nuclear power plant with two reactors. Shimane-1 and Shimane-2 have brought the largest subsidies to the prefecture. It is calculated that No. 2 reactor provided over 75 percent of the local tax revenue during the 1990s.[19] Shimane-1 was commissioned in 1974 and Shimane-2 was commissioned in 1989. Chugoku Electric Power Company owns and operates the plant, which is located in Kashima. In 2005, Kashima merged with Matsue, Shimane's capital, which puts the plant directly within the environs of a metropolitan area. Although in 2006 a seismic fault line was discovered close to the power plant, there were no plans to upgrade its structure.

The workforce's abiding reliance on public works projects and concerns among local authorities that the subsidies would ebb encouraged the request for the construction of a third reactor. The construction of the third reactor, designed by Hitachi, was underway in 2011. The commission date was set for March 2012. No. 3 reactor has brought local residents $90 million in public works funds and a guarantee for $690 million in property tax revenues to be allotted over a fifteen-year period when the reactor is commissioned. After the Fukushima Daiichi disaster, the residents were very worried about the safety of their nuclear facility but the need for a flow of government funds overwhelmed the concerns for safety and residents stoically accepted the risks.

Ehime Prefecture: the Takeshita connection

The elected officials in the National Diet who hold cabinet posts or who are in influential positions in a main political party can build enduring relationships with businesses and with the bureaucracy. Ehime Prefecture, the largest of four prefectures on Japan's fourth-largest island of Shikoku, was fortunate to have one of its representatives in the Lower House as the minister of construction in Prime Minister Takeshita's cabinet. LDP politician Ihei Ochi[20] was in the right place at the right time (1987–8) and as a consequence, the applications for public works in Ehime were accepted in the first Takeshita budget.

Governors also strive to create strong links to the national ministries. Local government budgets for entertaining ministry officials usually depend upon a prefecture's affluence. Former Ehime governor Sadayuki Iga took full advantage of Representative Ochi's connection with the MOC and the construction industry and during his twelve years in office until 1999 he cultivated an interpersonal network with top officials in the ministries to promote Ehime's needs.

Ehime has an area of 5,676.22 km^2 with 162.5 km of coastline, the fifth-longest in Japan. The population is approximately 1.5 million with 500,000 living in Matsuyama, the capital. The primary industries are agriculture (citrus), forestry and fisheries. Secondary industries are shipbuilding, chemicals, machinery and paper pulp. In terms of per capita GDP Ehime ranks 35th among the forty-seven prefectures, 32nd for personal income tax revenue, 25th in industrial output, 25th for agricultural produce and 30th for the size of the budget.[21]

Before he entered the governor's office, Iga had been an officer of the Ehime local government and a staunch member of the LDP. As vice-governor under the previous administration, he was in line to succeed as governor. His close ties to the ministries enabled him to access large subsidies for public works which included a network of highways and tunnels throughout Ehime's mountainous terrain, and the Kurushima Bridge that links Ehime with Hiroshima on the main island of Honshu and which was under the management of the Honshu–Shikoku Bridge Authority, one of the debt-ridden public corporations that was merged with the JH. Iga's efforts also brought an international airport, a convention center, a modern art museum and, his biggest triumph, the first Foreign Access Zone (FAZ) in Japan. By the end of his final term in office, Iga had gained a reputation among residents as a governor obsessed with self-glorification.

FAZ opened in the port city of Matsuyama in 1993. It was the largest project that Iga pulled into his prefecture. The former MITI initiated FAZ installations in response to the United States' demands that Japan open its markets to more imports. In 1992, MITI wrote the Law on Extraordinary Measures for the Promotion of Imports and the Facilitation of Foreign Direct Investment in Japan, setting the stage for the construction of Foreign Access Zones throughout Japan.

In January 1993 before the inauguration of FAZ in Matsuyama, MITI international vice-minister Yuji Tanahashi (Masahisa Naito's former colleague) met with foreign business executives in Tokyo to discuss the implementation of FAZ and MITI's plans to modify customs laws, create a fund for industrial structural adjustment, subsidize foreign investment, defray investment costs, guarantee loans, provide additional investment information and support employee recruitment. When asked if the difficulties experienced by foreign companies to access Japan's markets were due to lack of effort, Tanahashi replied that the Japanese language could be problematic but he pointed to the successful entry by large international firms such as Motorola and Texas Instruments.[22] Since 2008 Tanahashi has been serving as Chairman of the Japan Petroleum Exploration Co. Ltd., which prior to 2003 was an entity in JNOC.

In April 1993, a company comprised of government organizations and private corporations was established to manage FAZ. The Ehime Foreign Access Zone Co. Ltd. constructed a distribution center for the handling of imported goods and three exhibition halls, one of which was the largest exhibition hall in the Shikoku–Chugoku region. The hall is suitable for numerous events with conference rooms fitted with projection and simultaneous translation equipment. The entire exhibition area covers 7,300 square meters.

In the early 1990s, Iga's contacts in the ministries considered that the public works projects would be supported by Ehime's expanding economy and would support small business growth. A second FAZ was constructed in 1999. The project involved building a new port and roads connecting the facilities to the expressway system. The allocation of subsidies presented lucrative contracts to big business and provided employment for hundreds of local workers. Additionally, the ministries could expand influence. As an example, JETRO set up JETRO FAZ Support Centers in Tokyo and JETRO FAZ Support Centers in nine of the FAZ, including Ehime. The Tokyo Support Center distributes FAZ-specific information and data regarding each location. The center also offers advice on import and investment in Japan, consulting services with English-speaking staff who answer questions on trade and investment and how to search for and select partners. Since FDI in Japan is still one of the lowest among the OECD countries at 3.6 percent in 2010 it appears that the existing twenty-two FAZ were more relevant to bringing big construction contracts to big business and to extending ministerial guidance rather than opening up markets to foreign investment and imports.

It is difficult to determine how government intended to use all of the facilities over the long term. Ehime's economy was still deflated and local government coffers were nearly empty. The population was decreasing due to lower birth-rates and residents are moving to the more affluent prefectures to seek work. Furthermore, the expansion of industry in Ehime appeared unlikely because companies are relocating production facilities overseas to cut costs. FAZ is not being fully utilized as the facilities are often undersubscribed. Traffic is relatively light on the new motorways and commuters to Hiroshima prefer riding on the ferry to crossing the Kuroshima Bridge because the tolls are too expensive. The museum is a beautiful structure with luxurious mahogany toilets, but the small collection of art questions the justification of the expenditure. Also, it appears that no consideration was given to the ultimate cost of these projects, let alone the drain on local tax revenue for the maintenance of what are referred to as 'empty boxes'.

Iga was expected to win another term in office but the recession prompted dissatisfied voters to replace him in January 1999 with Moriyuki Kato, a high-ranking official in the Ministry of Education who retired from office to enter politics. The initial reaction from local business owners was positive because Kato's close links to the national ministries would hopefully serve to facilitate the procurement of public works contracts and subsidies. However, by Kato's second term in office, Ehime residents had come to realize that Kato's connections brought them more than they had bargained for.

Kato's connections to MEXT and METI: textbook and MOX

The controversial *New History Textbook* edited by Kato's former ministry was introduced to high schools in some prefectures in 2002. The governor, who was a proponent of the textbook, announced publicly in 2002 that he felt that the book 'is the most appropriate to deepen people's appreciation of the history of the country'. Although many local citizens were opposed to the textbook, Kato received more support from Koizumi and Abe on 15 May 2004 at an Ehime town meeting for the revised education bill.[23] The hall was packed with 100 educators with links to MEXT. A subsequent investigation of the event revealed the following November that sixty-five educators had been paid ¥5,000 each for attending the meeting and that one member was given the task of spouting statements in support of the textbook and asking questions prepared by the ministry. Despite a lawsuit filed in December 2005 against Kato by 1,000 people, including South Korean and Chinese, demanding the rejection of the textbook, it was distributed to schools in April 2006. Koizumi later admitted that he had arranged the rigged town meetings throughout Japan to promote government objectives.

Ehime residents also had not anticipated that the election of a former MEXT official would lead to a pluthermal reactor and MOX fuel. METI's JNC and MEXT's JAERI developed MOX and METI was determined to introduce the fuel in fast breeder reactors. In October 2006 when METI requested that Ehime host a pluthermal plant to become fully operational by 2010, in spite of vehement protests from civic groups, Kato welcomed the project since Ehime's economy would undoubtedly benefit from substantial government subsidies which would impact on domestic businesses and support employment. On 13 October 2006, after obtaining the consent from Ikata mayor Kazuhiko Yamashita, the governor presented a written agreement to the president of Shikoku Electric Power Company (YODEN), which owns and operates the power plant. Six months earlier Saga Prefecture in Kyushu had given permission to the Kyushu Electric Power Company (KYUDEN) to build a pluthermal reactor at its Genkai nuclear plant.

Ehime expected to receive ¥6 billion ($51.7 million) in subsidies. Kato expressed confidence about YODEN's safety measures and his belief that the central government's pluthermal policies were practical. Three years later, refusing to accept defeat, on 18 January 2009 the Association of Ehime Prefectural People for Cooperation as well as other citizen organizations requested that NISA force YODEN[24] to cancel its plans for the pluthermal reactor. Their demands were ignored and YODEN began commercial operations in March 2010. On 2 August 2011, *Mainichi Shimbun* reported that a retired METI official who headed the public relations in NISA admitted that NISA had asked YODEN senior officials to request that staff attend a seminar in Ikata in June 2006 where there was discussion of using MOX in Ikata's No. 3 reactor.

Ikata Nuclear Power Plant: money, money, money

The ¥6 billion payoff was a hefty amount, especially for a prefecture which generated only about 36 percent of its tax revenue. In 2010, Ehime's economy was dependent on the manufacture of parts for exports to emerging markets, principally China, Japan's biggest trading partner. The levels of sales for agricultural products and seafood remained flat and at about the same level as the previous year. The upswing in the demand for semiconductors, transportation and manufacturing machinery impacted favorably on Ehime's secondary industries but the production of pulp and paper, textiles and chemicals remained flat. The level of production of the other secondary industries was mixed and future production levels were difficult to forecast because of the slowdown of the global economic recovery and the instability of foreign markets.

As in the other prefectures, tax-breaks temporarily supported the sales of eco-friendly automobiles, white goods and home electronics but due to the severe conditions in employment and wages, sales in large and small retailers continued to fall. Poor conditions were expected to continue. Investment in the housing market and real estate was low. Government investment increased slightly but the increase was related to contracts that had been ordered during the previous year after the government had released a supplementary budget. The continuation of low capital investment and the 1.9 percent fall from the previous year of the average contracted long and short-term interest rates signaled that corporations were reluctant to expand operations.

The small town of Ikata is located on Japan's narrowest peninsula and the westernmost point of Shikoku. Its industry is limited to fishing and citrus farming. The mountainous terrain has prevented urban development but the substantial investment from the nuclear industry has supported infrastructure and tourism. Ikata Nuclear Power Plant is Shikoku's only nuclear installation and provides much of the island's electricity. No. 1 reactor, built by Mitsubishi Heavy Industries, began commercial operations on 30 September 1977. No. 2 reactor, also built by Mitsubishi, was commissioned on 19 March 1982. No. 3 reactor, which was partially loaded with MOX and began pluthermal operation in March 2010, was built jointly by Mitsubishi and Westinghouse.

At the age of seventy-six, Governor Kato decided not to campaign for a third term in office and to retire from politics. His successor was Torihiko Nakagawa, a local lawyer, who won the 2010 election on the LDP ticket. He also applauded the initiation of a pluthermal plant and agreed with Kato that the government's screening methods and the government nuclear policy were reliable.

The price of pork-barrel patronage

Tanaka's 'Plan for the Remodeling of the Japanese Archipelago' and the Land Development Policy did not stem the migration of the population from rural areas to the bigger cities and industrialized regions. Instead of a more devolved society, Tanaka's plans acted to engender more dependency on central government and on ministerial guidance. Moreover, Tanaka's pork-barrel system justified the plundering of FILP by politicians, ministry officials and big businesses. Regardless, while Japan's economy was expanding and tax revenues were feeding the treasury and Japan was overtaking the US economy, the Japanese preferred to ignore the spending spree.

In the 1990s, the government continuously released huge fiscal stimulus packages, allocating a large portion to public works. Hundreds of miles of roads, expressways, dams and bridges were built, providing construction companies with lucrative contracts and employment for thousands of workers while no consideration was given to how the new infrastructure would serve the local population in the future. Ultimately, infrastructure was built for the sake of building.

Before the bursting of the asset-inflated 'bubble', Japan did not have a public deficit. However, within the decade, Japan's sovereign debt escalated to 130 percent of the GDP. The downsizing of FILP and the privatization of the Postal Services signaled that the days of pork-barrel patronage were coming to an end. LDP politicians warned Koizumi that the party would suffer the consequences among rural voters who had enjoyed a steady diet of public works projects.

The local communities that accepted nuclear power plants and have been the beneficiaries of numerous public works projects are structurally dependent on the power industry and on central government to improve living standards, to provide employment and to procure from small businesses. Pork-barrel patronage paved the way for the entry of nuclear power plants and for the massive investments by power companies and heavy industries into these communities. The expenditure of public revenue on atomic powered energy served to finance the activities of the nuclear energy research and development organizations operated by METI and MEXT, perpetuated the nuclear energy industrial complex and made the host communities effectively the domains of Japan Inc.

The government regulation calling for the suspension of nuclear power plant operations to allow mandatory maintenance and safety inspection every thirteen months indicated that only nineteen plants would be operating simultaneously during the summer months resulting in power shortages. On 30 April 2011, Hokkaido Electric Power Company (HEPCO), KEPCO and CHUDEN announced that, after receiving approval from the local authorities, they planned to restart operations of the nuclear facilities that were idled after the earthquake for maintenance.[25]

Fukui Prefecture has thirteen reactors, six of which were idled for planned and unplanned maintenance. KEPCO, Japan's second largest electric power company, operates eleven. Fukui governor Issei Ishikawa announced on 13 June that he would not approve the restart of the offline reactors unless the government clarified the safety situation at all plants in Japan. The Oi Nuclear Power Plant, in Oi, is also operated by KEPCO and supplies much of the power to Osaka, Japan's third-largest city. One reactor had been idled for a regular inspection and the Oi authority was concerned that if the reactor was shut down permanently the subsidies that the town earned from hosting the reactor would stop. On 16 July, NISA reported that KEPCO was forced to manually shut down the reactor temporarily because of problems with the cooling system.

The town of Hamaoka, which merged with Omaizaki in 2004, was impoverished before the power plant was constructed. The town has received over $700 million in subsidies since the 1970s and the power plant employs 3,000 people. Although in favor of the development of renewable energy, the majority of residents wanted nuclear operations maintained.

On 15 June, Akihisa Mizuno, president of CHUDEN, reassured local residents that the company had taken safety measures at Hamaoka to counteract damage by earthquakes and that an accident that crippled Daiichi would never occur at Hamaoka. He emphasized that the three reactors operating at Hamaoka had been designed to withstand a major quake.

However, seismologists had warned the government in 1970 before the license for the construction of Hamaoka Nuclear Power Plant No. 1 reactor was granted to CHUDEN that the proposed site was near to two seismic fault lines and that a major earthquake in the Tokai region was overdue. Regardless, CHUDEN received the permit and commenced construction and the reactor was commissioned in 1976. Three more reactors were commissioned and, despite Professor Ishibashi's claim in 2004 that Hamaoka was the most dangerous plant in Japan, a fifth reactor was commissioned in 2005. No. 1 and 2 reactors were shut down in January 2009.

CHUDEN announced on 26 July that it expected a consolidated operating loss of ¥100 billion ($1.28 billion) in FY 2010. The company estimated that the substitution of thermal power to generate electricity would cost the utility ¥250 billion ($3.25 billion) in purchases of liquefied gas in FY 2011. In addition, CHUDEN needed ¥300 billion ($3.9 billion) to refinance bonds that would mature in FY 2011. In July the DBJ, Tokyo-Mitsubishi UFJ and Mizuho Corporate Bank loaned a total of ¥350 billion to the utility. In August, the Japan Bank for International Cooperation (JBIC) provided an emergency loan of ¥100 billion. Regulations require that JBIC loan to companies and institutions that are investing in the construction of infrastructure and resource development in other countries and not for increasing operating capital. In order to justify the loan to CHUDEN the bank provided the funds to cover costs of fuel imports from other countries.

The power of money

Nuclear power generates 30 percent of Japan's power supply. If operations were to be suspended for an extended period there would be serious power supply shortages and the consequences would be extremely serious for both large and small producers. Companies with major operations in the Osaka, Aichi and Gifu regions such as Toyota, Suzuki, Honda, Nintendo and Panasonic were pressuring KEPCO to reboot the idled plants as soon as possible. KEPCO, which depends on nuclear powered energy for about 50 percent of its power output, announced to shareholders at the annual shareholder meeting in June that the company's future focused on atomic power. KEPCO executives reassured investors that counter-measures had been established to deal with earthquakes and tsunamis that might damage KEPCO reactors operating in Fukui Prefecture.

The power companies and businesses were also pressuring the government to sanction the resumption of some of the idled reactors. The government announced on 7 June that the utility companies had taken correct emergency measures to ensure the safety of their power plants in case they were hit by a tsunami. The announcement did not include references to earthquakes. The government's report submitted to the IAEA the same day admitted that TEPCO had not prepared properly for the disaster, citing defects in the design of the reactors and problems with equipment. The report pointed to the tsunami as the cause of the plant's damage but it stated that an investigation regarding whether an earthquake had triggered the event was also being implemented.

TEPCO, which supplies 30 percent of electric power to the northeast region of Honshu, faced huge costs for compensation to local residents, farmers, fishermen and to workers who may suffer from the effects of radiation exposure and could face bankruptcy. However, TEPCO, the world's fourth-largest electric utility company with offices located in both Washington, DC and in London was too big to fail. It owed its creditors ¥3.4 trillion ($42 billion) with Sumitomo Mitsui Banking Corp. and other major lenders holding $27 billion worth of outstanding loans at the end of March. Kan's administration was keen to have TEPCO declare bankruptcy so that a newly established government-run agency could step in, purchase the equity and then restructure the company.

Since the 1950s TEPCO had close connections with the *keiretsu* and their banks and was a major procurer of supplies from a number of industries. A government-led bankruptcy would impact negatively on its long-term relationship with its suppliers and with their banks. Major financial institutions and insurers who assumed that TECPO shares were a stable asset reported a ¥400 billion ($4.94 billion) loss in FY 2010. No 1. Life Insurance Co. and Nippon Life Insurance Co. suffered losses of ¥100 billion ($989.36 million) each. Sumitomo Life Insurance Co. and Mitsui Life Insurance Co. reported losses of ¥10 billion ($123.6 million) and ¥5 billion respectively. Sumitomo Financial Group's loss was ¥80 billion ($989.36 million), Mizuho Financial Group lost ¥50 billion ($618.35 million) and Mitsubishi UFJ posted a ¥30 billion ($371.01 million) loss.

In order to secure government assistance with compensation payouts, TEPCO had to agree to not limiting compensation payments, streamlining operations and cutting costs and allowing a special agency to supervise the payouts. TEPCO announced on 26 April that, for the first time since 1951, it would not hire new staff (in February it had expected to hire 1,100 graduates and had received 6,500 applications) and that the salaries of the chairman, president, vice-president and managing directors would be cut by 50 percent. Annual wages of management staff would be cut by 25 percent and other employees would see a 20 percent reduction in their annual salaries. However, on 8 May Kan's cabinet ministers rejected the restructuring measures as inadequate and constructed a compensation package that ultimately put the burden of financing on taxpayers and consumers and not on TEPCO shareholders.[26]

In 2004, METI vice-ministers Katsusada Hirose and Seiji Murata had argued for the liberalization of TEPCO and METI officials, who had been under their patronage, were pushing to place the company under close government supervision before it went bankrupt and then restructure it. On the other hand, a rival faction of METI bureaucrats, who had been under the wings of former vice-ministers Hideji Sugiyama and Takeo Kitabata and who were in a number of top ministry posts, were in favor of keeping TEPCO as a private corporation. Furthermore, since its establishment, NISA had accepted eighty 'on-loan' staff from the power utilities and companies in the nuclear energy sector.[27]

The power industry has ties with LDP and DPJ politicians. Among them are Lower House member and former minister of economy, trade and industry, Akira Amari, in Fukuda's and Abe's administrations who, with other LDP politicians, is a member of the commerce lobby, and Upper House DPJ member Masashi Fujiwara, who has close links with the Federation of Electric Power Related Industry Worker's Union. Amari was opposed to separating NISA from METI.[28]

Despite Kan's intention to restructure the company through nationalization, METI officials and TEPCO's main banks managed to preserve the status quo, even though TEPCO would become essentially a 'zombie' company. Its main banks, Sumitomo Mitsui Bank Corp. along with Mitsubishi UFJ, Sumitomo Financial Group Inc. and Mizuho Financial Group Inc., had announced on 11 April a $25 billion loan to TEPCO. Mitsui and its affiliates also purchased shares of TEPCO to support the share prices.

On 13 May, TEPCO accepted the package as well as the unlimited cap on compensation claims.[29] The government provided about ¥5 trillion ($62 billion) for damages and created a body to handle the payouts for the claims against TEPCO, which according to the Bank of America-Merrill

Lynch could reach ¥11 trillion ($136 billion). TEPCO was expected to pay ¥200 billion ($2.47 billion) annually. Decommissioning Daiichi would cost a further ¥1.5 trillion ($18.57 billion) and the sum of ¥250 billion ($3.09 billion) was calculated to restore operations in other stricken plants. Along with tax revenue, the expected increase in electricity rates would counteract the losses of the major shareholders. TEPCO was scheduled to repay the government within thirty years which implied that, ultimately, the public would bear the brunt of TEPCO's debt. The government asked TEPCO's creditors to waive TEPCO's debts but the banks refused because the debts were so large a waiver would cause instability in the finance industry. Indeed, TEPCO's junk-bond status was destabilizing the entire corporate bond market. The TEPCO compensation bill was formally approved by the LDP, New Komeito Party and the DPJ and passed by a Lower House committee on 26 July.

Following TEPCO's announcement, Moody's Investor Service downgraded TEPCO's long-term bonds from Baa1 to Baa3, one notch above junk bonds. On 30 May for the fourth time since 11 March, Standard & Poor's lowered TEPCO's rating from BBB to B+ or junk status. The ratings agency stated that TEPCO's creditors may be required to write off TEPCO's debt of nearly ¥4 trillion ($49.38 billion). On 20 June, Moody's also rated TEPCO's long-term issues as junk (i.e. B1). On 23 June, TEPCO reported an extra $1.5 billion loss for compensation payouts and an additional $473 million for dealing with the ongoing crisis at Daiichi.

On 28 June, 9,302 TEPCO shareholders, the largest number on record, met at the annual shareholders meeting to vote on proposals put forward by TEPCO and by fellow shareholders. Disregarding the fact that the crisis at Fukushima had wiped $36 billion of TEPCO's market value leading to TEPCO's junk-bond rating, an overwhelming majority voted against a proposal submitted by 402 shareholders requesting that the company abandon the nuclear power business. They also rejected a request to dismiss Chairman Tsunehisa Katsunuma.

On 2 July, Moody's also downgraded the eight other electric power utilities, including CHUDEN, KEPCO and KYUDEN to A1 and Aa2. Regardless, the majority of shareholders supported the continuation of nuclear power.

The power of the state

On 22 May, METI issued an in-house document stating that it would seek to keep nuclear power plants and nuclear power generation as one of three key power supplies and that it planned to set up a new panel under its Advisory Committee for Natural Resources and Energy, the organization that had developed Japan's energy policies.

On 10 June, Banri Kaida announced at an informal ministerial session that if the utilities did not reboot their power plants after safety inspections the overall economy would suffer, especially in western Japan where companies were stepping up production to compensate for the slowdown in northeastern Japan. He was very concerned that the continuation of power shortages would force the large Japanese producers to move more production facilities abroad.

On 18 June, NISA announced that all short-term measures to prevent severe accidents had been taken at plants which had been idled for inspection before the earthquake. Kaida followed by stating that eleven of the plants were deemed safe to restart and that he would visit local governments hosting nuclear power plants to gain the acceptance of officials and residents. Kan agreed to the resumption of operations if the plants met government safety regulations.

On 19 June, Haruki Madarame, chairman of the NSC, acknowledged publicly that the commission's role in monitoring Japan's nuclear policy was rigid, that it would be reassessed and that, although there would probably be changes to the monitoring system, the NSC would always be involved.

On 26 June, to counteract Genkai residents' opposition to restarting the No. 2 and No. 3 reactors that had been idled for safety checks METI broadcast on local television and the Internet reassurances to local residents that the reactors were safe to resume operations. The acceptance by the Genkai local government and by the Kyushu Prefectural government of the resumption of Genkai's reactors was crucial because it would pressure other prefectures to allow the resumption of operations at the other idled power plants throughout Japan. It would also impact favorably on national policy to increase the number of pluthermal power plants and MOX fuel.

On 29 June, Banri Kaida visited Genkai to meet with Saga Governor Yasushi Furukawa and the mayor of Genkai Hideo Kishimoto to request that local government accept the resumption of the two reactors that had been idled for regular safety checks. Furukawa was the first governor to be asked by central government to give permission for the resumption of the reactors in Genkai. Kishimoto consented to KYUDEN's[30] request initially. Furukawa was about to give his consent but withheld his final decision because of the mixed messages from the government about the new safety standards.

It appeared that the majority of the idled reactors would go online again in July but Kan, without consulting Kaida or his cabinet members, abruptly announced the following week his plans for creating a nuclear-free society and shifting government policy from nuclear to renewable energy. He called for a new series of stress tests to be conducted on all of the nuclear reactors. Even though Kan later apologized for causing confusion after his government had committed to the safety of the power plants and acknowledging that the plan to shift from nuclear to renewable energy was a personal plan and not a government policy, the damage was done. Kaida threatened to resign.

Genkai mayor Kishimoto, both humiliated and infuriated by Kan's sudden decision and the government's inability to coordinate its policy on nuclear safety standards, quickly withdrew his acceptance of Genkai's restart. Other prefectures followed suit. Ehime governor Nakagawa withdrew his support for the restart of Ikata until the government could guarantee the safety conditions at Ikata.

Governor Hirohiko Izumida also refused to permit TEPCO to restart Kashiwa-Kariwa Plant's No. 2–4 reactors that had been undergoing standard safety checks. Izumada told the press on 26 July that the new stress tests were useless unless the cause of the accident at Daiichi was understood. The suspension of No. 7 reactor in August for an 83-day scheduled checkup left only No. 5 and 6 reactors operating.

An additional set-back to the restart of Genkai was the revelation by a whistle-blower inside KYUDEN that KYUDEN executives in late June, in order to support METI's televised broadcasts on 26 June to Genkai residents, had used staff to masquerade as ordinary citizens and launch an email campaign posting positive comments about GENKAI's safety on the Internet. METI apologized for the incident and Kaida pressed KYUDEN's president to restore the public's trust. But at the end of July, Governor Furukawa confessed that he had suggested to KYUDEN's former vice-president that the company urge staff to send emails to solicit public support in order to increase KYUDEN's support to restart Genkai. Nevertheless, KYUDEN, stinging from the email scandal, warned that if Genkai No. 2 and No. 1 reactors at its Sendai plant in Kagoshima Prefecture were not back online, its power supply ratio would drop 4 percent from 18.5 percent during the summer peak season.

KYUDEN was almost certain to receive the go-ahead to restart No. 3 reactor at Genkai when it discovered that safety inspection data issued in 2009 by Obayashi Corp., a leading general contractor, were flawed and would have to be repeated, leading to a three-month delay. Also,

Obayashi Corp. had inspected eight other plants in Japan and it was necessary to repeat those surveys as well.

At a news conference on 11 July Minister Edano told reporters that the stress tests would not take nearly as long to conduct as the stress tests being conducted on European reactors, which were predicted to take seven months. It was envisioned that the idled reactors would be online within two to three months and that the entire procedure would take six months. Edano also cautioned that the shutdown of the reactor at KEPCO's Oi plant would affect power supplies throughout western Japan and requested that its customers conserve use.

The new stress test was modeled on the new European rules and was a two-step affair. After the first step was completed, the reactors could go online. The power companies would evaluate their power plants' ability to withstand earthquakes and tsunamis, NISA would then inspect the results of the tests and the NSC would evaluate the assessments. The NSC was reticent to be directly involved in the decision to restart the power plants because as an organization established under the Cabinet Office, its responsibility was to give recommendations to the prime minister and to government organizations. After obtaining the approval from the NSC on 22 July, NISA announced that it would begin the stress tests. However, since Kan had acted unilaterally without considering the legal implications and since the local governments could authorize the restart of plants independently, the exercise did not solve existing problems and served only to delay Japan's post-disaster recovery and promoted the relocation of manufacturers operations to other countries.

Edano claimed that the new stress tests should restore public trust in nuclear power because the NSC operated independent from NISA and as a 'third party' could be trusted to confirm NISA's results.

In 2007, the IAEA[31] reiterated Professor Ishibashi's views that NISA and the NSC were intertwined and that there was no effort to define the roles of each organization. It recommended that the role of both organizations be 'clarified' particularly in regards to setting safety guidelines. The IAEA also questioned the independence of NISA from METI even though it referred to itself as a 'special institution' and independent from METI. However, the director-general Nobuaki Terasaka was a METI bureaucrat and the 2011 METI vice-minister Kazuo Matsunaga was a former NISA chief.[32]

On 12 July, Keidanren urged the government to restart the reactors that had been idled for safety checks and, also, to 'steadily promote' nuclear power after it had restored the public's trust in its energy policies.[33]

On 7 August, Hokkaido governor Harumi Takahashi, a former METI official, after discussion with local authorities announced her approval of the resumption of commercial operations of the Tomari No. 3 reactor which is operated by the Hokkaido Electric Power Co. (HEPCO). Tomari is a fishing village with a population of 2,000. Fifty percent of its revenue is generated by nuclear-related money and HEPCO donations. It is the only town in Hokkaido that hosts a nuclear power plant and, contrary to the other villages in Hokkaido, its economy is in surplus. The Tomari plant had been undergoing regular inspection in January and a readjustment on 7 March. However, commercial operations were postponed due to the nuclear crisis. Tomari was the first reactor to receive local government approval, forging the way for other local governments to follow suit.

The government's post-11 March national energy policy specifies a gradual reduction in the number of nuclear power plants and an investment by government equal to 1 percent of GDP for developing renewable energy sources. Nevertheless, for eighteen months the DPJ-led government had aggressively promoted nuclear technologies to foreign markets. METI arranges trade missions abroad to promote business opportunities for Japanese companies. In April 2009, Yukio Hatoyama met with the Vietnamese trade minister to discuss the formation of a strategic partnership to introduce nuclear power to Vietnam. He also met with Masayuki Naoshima, Hatoyama's METI minister to discuss the collaboration. In May, METI officials and executives from Toshiba, Hitachi, Mitsubishi Heavy Industries, TEPCO, KEPCO and CHUDEN visited Vietnam for meetings with Vietnamese government officials and industry. Prime Minister Kan played an instrumental role in the signing of the nuclear power agreement with Vietnam. In January he stated in a policy speech in the Diet that he had worked personally on the project and that the export of a nuclear power plant had been successful.

On 23 August the government announced that it would resume its engagement with industry to export nuclear energy technologies, which had stalled after 11 March. Anticipating that emerging economies would embrace nuclear power generation to meet their growing demands for electricity, METI estimated in 2010 that by 2020 the global market for nuclear power would be worth ¥16 trillion ($207 billion). Inevitably, nuclear power will remain big business for Japan's nuclear village.

On12 September at his first policy speech before the Diet, Prime Minister Noda called for the restart of the nuclear plants. The following week, speaking to UN delegates about Fukushima Daiichi, he promised a cold shotdown by year end and the resumption of the plants undergoing safety checks by the spring. He also promoted the export of nuclear reactors and technologies.

7

Japan's Nuclear Crisis: the Routes to Responsibility

Japan's nuclear energy program began in 1954 with the objective of developing an energy source that supplied the country's growing need for electricity. The Atomic Energy Basic Law, proposed by Prime Minister Yasujiro Nakasone, was enacted in 1955 which called for independent, democratic and transparent activities in nuclear research. Former Lower House representative Matsutaro Shoriki also played a major role in the promotion of nuclear energy. Shoriki, who was also the owner of the *Yomiuri Shimbun*, was the first chairman of the Atomic Energy Commission, which was founded in 1956 to promote nuclear energy power development and the use of nuclear fuel. Nakasone succeeded Shiori as the second chairman. The United States fully supported the introduction of nuclear in Japan as an integral part of its global strategy.

At the same time, the Science and Technology Agency, the Japan Atomic Energy Research Institute (JAERI) and the Atomic Fuel Corporation were established. In 1957, the Japan Atomic Power Company (JAPCO) was formed to spur on the commercial use of nuclear power. JAPCO[1] is owned by Japan's electric power companies: TEPCO, KEPCO, CHUDEN, Hokuriko Electric Power Co. and the Electric Power Development Company (EPDC KK), which was renamed J-POWER in 2005. As a division in MITI, EPDC oversaw the electric power companies.

In 1959, MITI established the Japan Nuclear Fuel Corporation, the forerunner of the Japan Nuclear Cycle Development Institute (JNC). In 1967, MITI established the Power Reactor and Nuclear Development Corporation (PNC) as the major fuel cycle research institute in Japan with the research center in O-rai, Ibaraki Prefecture. The PNC engaged in numerous activities, including uranium exploration, high-level nuclear waste disposal, spent fuel reprocessing, MOX fuel fabrication and the development of a fast breeder reactor (FBR) for the use of MOX.

However, when the PNC was unable to respond to several accidents such as the one at Monju in Fukui Prefecture MITI reorganized it within the JNC.

The government's long-term nuclear energy policy targeted nuclear power as the major provider of electrical power and in 2011 it accounted for 30 percent of Japan's electricity. The policy also targeted the recycling of uranium and plutonium from spent fuel for light water reactors (LWRs), the development of FBRs, and the promotion of nuclear energy to the public with the pledge that it was safe and that nuclear power plants would be used only for ensuring a stable supply of energy and not for the production of nuclear weapons.

The JNC focused on the development of FBRs, reprocessing spent fuel, MOX fuel fabrication and high-level waste disposal. The JNC was merged with JAERI in 2005 to create the Japan Atomic Energy Agency (JAEA).[2] The agency, which includes a number of energy R&D institutes, has 4,400 staff, ten branch facilities and an annual budget of approximately $1.7 billion.

In June 2002, a new Energy Policy Law was enacted that gave the government greater authority to establish energy infrastructure in order to guarantee a stable energy supply for economic growth. Notwithstanding the data falsification scandal in 2002, the government forged ahead to increase the role of nuclear power in Japan's electricity needs. In May 2006, METI published the 'New Energy Strategy', a paper which defined METI's objectives for reducing Japan's reliance on fossil fuels, developing overseas natural resources and maintaining or increasing the ratio of nuclear power for all power production to 40 percent by 2030.

Japan's first commercial gas-cooled nuclear reactor was imported from the UK and began operating in 1966 in Tokai-mura, in Ibaraki Prefecture. It continued to operate until 1998. The nuclear reactors that followed were LWRs which used enriched uranium. In 1970, three LWRs began commercial operations. At first, Japanese companies joined with American companies, which possessed the know-how in the construction of reactors, but by the end of the 1970s Japan's heavy industrial companies such as Hitachi Co. Ltd., Toshiba Ltd. and Mitsubishi Heavy Industries Co. Ltd. were able to design and build plants independently for the domestic market and eventually export their services to East Asian markets.

The Fukushima Daiichi disaster was the most serious in a series of accidents to have occurred at Japan's nuclear power plants. But other events that have also plagued Japan's nuclear power industry since the early 1990s have caused deep anxieties among the residents in the towns that host the plants about the safety of the nuclear facilities. Fukui Prefecture, which is located on the Sea of Japan in western Japan, has thirteen commercial nuclear power plants and some of them are targeted to carry MOX fuel, despite the protests from local residents. The town of Tsuruga hosts several experimental FBRs using MOX fuel developed by the JNC. The JAEA has a head office in the town along with a Fast Breeder Research and Development Center and the FUGEN Decommissioning Engineering Center.

JOYO and FUGEN: trial and error

In April 1977, the PNC began testing its experimental FBR, JOYO, Japan's first liquid metal FBR.[3] The prototype's design was based on the research conducted at PNC's research center in O-rai. In June 1983, the successful reprocessing of spent fuel at JOYO was achieved. The completion of an FBR nuclear fuel cycle paved the way for the reprocessed fuel to be utilized in FUGEN, PNC's prototype FBR in Tsuruga.

On 20 March 1978, FUGEN began its test run. Built in the Tsuruga Nuclear Plant complex by the PNC, the FBR began full operations a year later. FUGEN was Japan's first domestically designed power plant and the world's first reactor to use a MOX fuel core. However, a serious accident occurred, which indicated that the reactor's design was flawed. On 14–16 April 1997, PNC officials postponed for thirty hours reporting a tritium leak. An investigation later revealed that eleven such episodes had occurred, resulting in the resignation of five PNC managers. The cover-up prompted MITI to restructure the PNC and integrate it in the JNC. When a giant spew of steam escaped from a broken pipe, the reactor was shut down. During the dismantling process a number of weak points were discovered in the reactor's walls.

MONJU: more trials and errors

In April 1994, PNC began testing MONJU (named for a Buddhist god), its second prototype FBR in Tsuruga. It is sodium-cooled and loaded with MOX fuel. On 8 December 1999, a defective pipe, carrying liquid sodium, burst and temperatures in the room in the reactor climbed, producing a heat so intense that some of the steel structures melted. A fire broke out due to a broken sensor and the JNC was forced to shut down the reactor. Although the leak was not radioactive, the public was angered by the JNC's attempt to cover up the seriousness of the accident and the extent of the damage.

The ensuing inspections of MONJU showed that the design of the plant was also flawed and that safety procedures had been violated. After four years of inspections, which included seismic safety checks and a review of modifications of the reactor, JNC submitted an application to METI for a license to modify MONJU. When the license was granted in December 2002, JNC stated on its MONJU website that it 'intended to start plant modification as soon as we obtain the understanding of the local community'.[4] However, it took years before the MONJU plant would go online again.

On 24 November 2000, when the JAEA, the plant's operator, announced that it planned to restart MONJU, Fukui residents and anti-nuclear activists filed a plea with the High Court to obstruct JAEA's plans. On 27 January 2003, the Nagoya High Court ruled to reverse its 1983 approval for the construction of MONJU because of the poor construction and because no consideration had been given regarding the magnitude of the dangers involved in the use of liquid sodium coolant. Nevertheless, JAEA, MONJU's operator, and the JNC, MONJU's builder, were victorious in the end. Prime Minister Koizumi, who was also determined to push the development of FBRs, throughout his administration promoted nuclear power as an absolute necessity for a stable energy supply. The following June, Koizumi emphatically told the Diet that the FBR and MOX program must continue and that the termination of MONJU was not an option. On 30 May 2005, the Japan Supreme Court gave JAEA permission to begin operating its plant.

It took five more years for the JAEA to convince Tsuruga residents and the Fukui governor that MONJU was safe. The JAEA held over 10,000 meetings with residents living in and around Tsuruga and conducted guided tours for an estimated 40,000 people.

On 6 May 2010, after receiving the final approval from the local authorities and the go-ahead from NISA, the JAEA announced that the plant would begin test operations on 8 May and supply some power to the region's grid. However, the startup was delayed when a device used to load the fuel fell inside the reactor vessel, and since it was soaked with sodium coolant and could easily ignite if it made contact with air the retrieval took months. Finally, in June 2011, the all-clear to start operations was given by the JAEA. By 2014, the reactor was expected to be fully operational. Fukui residents were not convinced that MONJU could withstand a major earthquake especially since seismic fault lines were discovered within 40 km of the reactor. There were concerns that the JAEA was underestimating the potential dangers because MONJU had been designed according to earthquake standards that were set in the 1980s.

The government continued to pressure the local government to accept the construction of two more reactors in the Tsuruga plant. On 10 June 2002, Cabinet Minister of Economy, Trade and Industry Takeo Hiranuma visited Fukui governor Yukio Kurita and Tsuruga mayor Kazuharu Kawase to urge them to agree to allow JAPCO to build two new reactors in the Tsuruga power plant. Kurita, who had been opposed to the construction, gave JAPCO permission on the condition that Japan's oldest reactor which JAPCO had built in 1970 was shut down and dismantled. JAPCO agreed. However, at the end of 2006, JAPCO announced that, due to a two-year delay in the construction of two new reactors and due to the government's revision of earthquake-resistant guidelines and construction regulations, it was postponing the shutdown of the old reactor. Mayor Kawase accommodated the extension of operations of the old 1970 reactor because Tsuruga's industrial structure depended on the nuclear power industry.

On May 2011, radioactivity spiked in No. 2 reactor, signaling the possibility of a damaged fuel core. JAPCO, which had admitted the month before that its No. 2 reactor did not have a back-up cooling system, asserted that the radioactivity was within normal limits. The reactor was shut down for inspection of the emergency cooling unit and the fuel assemblies in the core.

Mayor Kawase and Governor Kurita were unfazed by the event and on 17 June, they requested that JAPCO's planned construction of the two new reactors, No. 3 and No. 4, continue on schedule.

Mihama: still more errors

Mihama is also located in Fukui Prefecture on the coast of the Sea of Japan. With a population of approximately 11,000, Mihama hosts the Mihama Nuclear Power Plant, which is owned and operated by KEPCO. Mihama has three reactors. No. 1 began commercial operations in 1970, the second started operating in 1972 and No. 3 reactor was commissioned in 1976. The power plant has the dubious reputation for experiencing more serious accidents than any of the other nuclear installations in Japan. On 2 September 1991, a small amount of radiation was released into the atmosphere when a tube broke. On 17 May 2003, two holes opened simultaneously in No. 2 reactor's generators but no radiation was released.

On 9 August 2004, Japan's most serious accident to date occurred when four workers were killed and seven injured by a scalding hot steam leak in No. 3 reactor, which was built by Westinghouse. The event occurred when the plant was about to undergo routine maintenance. Safety regulations call for pipes to be checked for corrosion and to be replaced if the walls are worn and too thin. KEPCO had replaced the original pipes in only one reactor. The pipes in No. 3 reactor had been neither checked nor replaced. Nine months prior to the maintenance check, a subcontractor had informed KEPCO that the pipes should be inspected but KEPCO ignored the warning. The pipe walls were so thin that they burst, releasing the steam. NISA ordered KEPCO to suspend operations and perform a complete inspection. It also directed the other electric power companies to carry out similar inspections on their power plants.

Takahama: more errors but more MOX

KEPCO's nuclear power plant in Takahama, Fukui Prefecture houses four reactors, designed by Westinghouse and commissioned in 1974, 1975 and 1985. On 14 May 2010, 2.5 liters of coolant leaked from one of the reactors to form a solid mass on the floor of the container vessel. Nonetheless, on 23 December 2010, KEPCO announced that MOX fuel was being loaded into one of the reactors and that the reactor would begin generating electricity within a few days. The reactor was the fourth reactor in Japan to be loaded with MOX and to enter pluthermal operations.

Tokai-mura

Tokai-mura is located on the coast of the Pacific Ocean in Ibaraki Prefecture, which is about 75 miles northeast of Tokyo. The JAEA has a head office in Tokai as well as the Tokai Research and Development Center previously known as the Nuclear Fuel Cycle Engineering Laboratories, which was managed by JNC. The JAEA operates the O-rai Research and Development Center, which is also engaged in nuclear fuel engineering (JNC).

Tokai-mura hosted Japan's first nuclear power plant. Operated by the JAEA, the plant housed two reactors, the first beginning commercial operations in 1966 and the second in 1978. The first reactor was closed in 1998. The second continues to operate commercially.

The worst nuclear event in Japan to occur before the Fukushima Daiichi crisis was at the Tokai-mura Fuel Reprocessing Plant that was operated by Japan Nuclear Fuel Conversion Co. (JCO), a subsidiary of Sumitomo Mining Metals Co. The plant, an experimental facility under the JNC management, was not engaged in commercial operations. Until the criticality accident on 30 September 1999, the plant was at the center of Japan's spent fuel reprocessing operations. The plant was commissioned in 1988 and reprocessed uranium solution taken from a JCO fuel plant to be used in the fabrication of MOX fuel which was targeted for the experimental JOYO. The reprocessing plant was recycling over 3 metric tonnes of uranium per year, which was much higher than for normal reactors.

An explosion and fire at a nuclear waste storage processing plant had occurred two years earlier when thirty-seven workers were exposed to high levels of radiation. Some radiation spread to Tokyo and to nearby prefectures. Operations at the facility were temporarily suspended until 2000 when in November the Ibaraki Prefecture and Tokai-mura governments granted the resumption of operations provided that new safety regulations were adopted. However, civic groups in the area opposed the decision, claiming that the authorities had not made public the results of the analysis of the cause of the event and that reprocessing would destroy the evidence.

The September 1999 criticality accident occurred at the JCO Conversion Test Building on the morning of 30 September 1999. A nuclear chain reaction was triggered by three workers who were converting enriched uranium into a small batch of oxide powder for MOX fuel to be used in JOYO. The event continued for eighteen hours. Unknown to the state regulators, JCO's work procedures had been modified three years earlier to allow uranium oxide to be dissolved in stainless steel buckets instead of in a dissolution tank. Also, to save time, workers poured the solution directly into the precipitation tank, ignoring the normal mixing process. The workers did not adhere to standard safety procedures set by government and they were not properly equipped with protective suits or control systems. The state regulators visited the plant twice annually but never when it was operating.

According to JNC officials, the workers had poured into the precipitation tank uranium compound that was eight times more than the critically safe limit together with nitric acid solution. There was a blue flash and the room filled with a blue haze. The workers began feeling ill. The NSC assessed the situation as a criticality accident.[5] The government declared an evacuation of the area around the complex, cautioning residents to remain indoors until the following day. The JAEA announced later that afternoon that radiation levels around the plant had spiked to 10,000–20,000 times above normal levels and altogether forty-nine people were contaminated. Three workers received well over the lethal radiation dose of 8,000 mSv and were hospitalized. One worker died after twelve weeks, another worker succumbed twelve weeks later and the third worker died seven months after that.

The IAEA reported that the accident was caused by human error and a violation of safety regulations.[6]

Small and medium-size businesses whose operations were affected by the accident suffered big losses. Agricultural, marine and livestock products produced in the Tokai-mura area disappeared from retailers and supermarkets due to rumors that the foods contained radioactive substances. The Law on Compensation for Nuclear Damage limited the total amount of compensation for damages from an insurance company at ¥30 billion for a nuclear power plant accident. The claims for damages increased to over ¥6 billion and JCO offered to pay out over ¥1 billion for legitimate claims. Criminal charges were filed against JCO.

The blame game

The government, the utility companies and the regulator were confronted by an angry public for their slow response to the Fukushima crisis and for the inconsistent and sporadic information regarding the seriousness of the nuclear meltdown at Daiichi. Public opinion polls taken in April and May showed that the majority of Japanese no longer trusted the information released by either the government or TEPCO. The public was also frustrated with TEPCO's dilatory approach in dealing the crisis.

Toshiso Kosako, professor of radiation safety at Tokyo University, resigned as one of Prime Minister Kan's six senior advisors on 29 April when he lashed out at the government for disregarding the radiation safety limits for schools located in the vicinity of Daiichi, for ignoring the laws for the management of nuclear crises, and for improvising measures that were prolonging the crisis. He condemned the government for a lack of transparency concerning the dangerous levels of radiation. On 27 April, two days before his resignation Kodama submitted a report to the prime minister's office in which he criticized Kan for his lack of leadership as well as the NSC's reticence to cooperate with Kosako's team. The report claimed that the government had delayed the forecast of the proliferation of radiation from Daiichi which had been compiled by the Nuclear Safety Technology Center's data collection system SPEEDI (System for the Prediction of Environmental Emergency Dose Information).

Kosako's report accused the government of not using the advisors' technical know-how effectively and of not examining populations in Fukushima and neighboring prefectures for radiation contamination because of the risks of thyroid cancer in children. Kodama's abrupt departure served to exacerbate the criticism of Kan's handling of the nuclear crisis.

On 29 April, Kyodo News released a survey taken of forty-six prefecture governors conducted from mid-April, showing that twenty-five governors disapproved of the way that Kan had handled the earthquake–tsunami–nuclear disaster. Thirty governors complained about the way that the government had released information, particularly the severity of the nuclear crisis. Nine governors were not in favor of expanding operations at existing complexes while five governors stated that they would refuse all new reactors. Some of the governors advocated new government bond issuances to cover the costs of the triple-disaster, the cancellation of the cuts in public spending as well as the raising of the consumption tax[7] which the DPJ had pledged during its election campaign.

On 1 May, Hiromasa Yonekura, the chairman of Keidanren and also the chairman of Sumitomo Chemical Co., berated the government for placing the blame for the nuclear crisis on TEPCO. At an interview in Paris with Agence France Presse (AFP) Yonekura criticized officials for not taking full responsibility for the disaster. He argued that the magnitude of the disaster could not have been foreseen, that TEPCO had built the plant according to safety regulations and that TEPCO's response had been appropriate. Yonekura referred to the Nuclear Power Generation Compensation Law which states that power companies should not be forced to assume compensation for disasters that cannot be anticipated. He also claimed that the compensation deal with TEPCO was unreasonable and that compensation should come from a sales tax. Keidanren supported a gradual rise of the tax to 10 percent and the proposal that the tax should not only fund the social security system but also be applied to earthquake relief.

At a news conference in Sendai on 15 July, Yasuchika Hasegawa, the chairman of the Japan Association of Corporate Executives, also blamed Kan for his failure to give a clear plan for a shift from nuclear energy to renewable energy. He told reporters that Kan's sudden announcement gave the impression that renewable energy was achievable within a short period of time but that was completely unrealistic. Hasegawa was in favor of renewable energy but there were a number of issues to be considered such as the amount of time and money it would take to achieve a nuclear-free society:

> The government has a responsibility to explain medium and long-term strategies that can convince the Japanese people and business community. That the government is not meeting its responsibility is extremely displeasing and regrettable.

Hasegawa felt that Kan was an obstacle to rebuilding Tohoku and to revitalizing the economy.[8] Politicians and businessmen agreed that Kan was delaying Japan's post-crisis recovery.

Kan not only mistrusted bureaucrats, he also did not trust the 1986 crisis management system, which gave more power to the prime minister's office in times of a crisis. But the scheme called for the executive to collaborate with officials from the various ministries. Kan preferred relying on close associates to advise him which effectively isolated him from the institutions that collected pertinent data regarding the level of radioactive contamination from the reactors.

Although Kan acknowledged that the state was responsible in part for the nuclear disaster, he put the blame on TEPCO because it had not taken sufficient measures to protect Daiichi against a tsunami and because TEPCO was poorly prepared to deal with the crisis. TEPCO responded that according to the Nuclear Damage Compensation Law the company should be exempt from bearing the compensation damages because the tsunami that hit Fukushima was a 'natural disaster of an exceptional character'. Some DPJ officials sided with TEPCO.

Although the power utilities have been linked closely with the LDP and have contributed millions of yen to LDP politicians, first through direct contributions from the companies until 1976 when it was formally curtailed, and then through individual contributions from the companies' top executives, DPJ politicians also have interpersonal relationships with the electric power companies as former employees and through the acceptance of campaign contributions. Masashi Fujiwara, a DPJ member in the Upper House, who was employed as an electrical engineer at KEPCO and who served as secretary-general of the company's trade union, said that since 2007 his party considered nuclear power to be fundamental to a stable energy supply.

The *Japan Press Weekly*, published by the Japan Communist Party, claimed on 5 June 2011 that TEPCO's union had established a political league in 1976 specifically for the collection of campaign funds for politicians, and that 112 DPJ members in the Lower House were supported by TEPCO's union. According to the weekly, Fujiwara had received ¥36 million in campaign contributions from TEPCO during 2007–10. The article referred to a session of the Upper House on 28 April when Fujiwara criticized Kan for making TEPCO primarily responsible for the compensation claims and duping the media and the public into thinking that TEPCO was 'evil'.[9]

TEPCO's labor union belongs to the Federation of Electric Power Related Industry Workers' Union of Japan. The federation's political action committee contributes funds to the DPJ. The *Asahi News* reported that DPJ Upper House member Masao Kobayashi, a one-time member of TEPCO's labor union, received ¥30 million.[10]

TEPCO's corporate website includes a 'corporate ethics and compliance section' installed after the data falsification scandal was reported in 2002. TEPCO pledged that it had implemented measures to renew public confidence in TEPCO's safety inspection and maintenance operations of nuclear power facilities. TEPCO promised customers that it would focus efforts to ensure compliance with safety standards, transparency of its operations and the release of information. TEPCO also pledged to set up an internal safety inspection system and a scheme that would allow a 'third party' to check nuclear safety and quality control.[11] However, the promises were superficial if taken in the context of TEPCO's denial of Daiichi's vulnerability to earthquakes and tsunami and the failure to prepare Daiichi for the eventuality of a major earthquake.

The public's anger regarding the central government's indecisiveness in the handling of the nuclear crisis, the delay in the release of pertinent information concerning the levels of the radioactivity and the shoddy supervision of nuclear power plants by the nuclear regulatory body NISA impelled Kan to go on the offensive. Kan also placed the blame on NISA for the lax supervision of Daiichi, on the vested interests of NISA, the safety regulator of nuclear power plants, and METI, the promoter and developer of nuclear energy, and on collusion between METI officials and the electric power companies.

Splitting NISA from METI: the image of reform

NISA was established on 6 January 2001 as a part of the reorganization of the central ministries. The nuclear safety regulatory apparatus concerning the use of nuclear energy that had been managed by the Science and Technology Agency, which was established in 1956, was reorganized as a new division in MEXT. The Science and Technology Agency's responsibilities had involved the regulation of testing and research reactors, nuclear fuel facilities, radioactive waste management and R&D.

NISA's website promotes itself as 'an institution dedicated to ensuring nuclear power safety'. NISA declares that it is dedicated to 'making a sustained effort towards on-site safety' and assures the public that it is committed to keeping the public informed. NISA claims that first and foremost its mission is to recognize the risks that are intrinsic in nuclear installations and through regulatory measures to ensure the safety of the public. NISA also maintains that its objective is to respond quickly to accidents in order to control damage and the acceleration of events.[12]

Kan told the JAEA at a meeting on 7 June that his administration recognized the need to revise the nuclear regulatory system in order to regain international trust in Japan. He said that the government was considering the separation of the NISA from METI in order to guarantee transparent and unbiased inspections of nuclear power plants. The government initially proposed the unification of the NSC and NISA which would take over the monitoring of radioactive contamination from the Science and Technology Agency. The government also wanted to remove the responsibilities of nuclear regulation from METI and the NSC because of the interpersonal networks. As an example, Atsuyuki Suzuki, the former chairman of the NSC, was a regulator at MONJU. In 2010, he was appointed the head of the JAEA, the operator of MONJU.[13]

Forming an effective regulatory system involving METI was considered problematic because of the conflict of interests with power companies where METI officials found post-retirement positions. After much deliberation, the government decided in August to put the new Nuclear Safety Agency in the Ministry of the Environment. However, simply transferring the duties to another ministry in order to guarantee a transparent and unbiased inspection of power plants may prove to be a useless exercise for the following reasons:

1. Maintaining independence from the Ministry of Environment is questionable due to vested interests related to the ministry's operations.
2. MEXT will still be in charge of inspecting nuclear facilities and monitoring nuclear substances to prevent their use for nuclear weapons. MEXT's JAERA and METI have interests in the development of nuclear.
3. Although the environment minister will control setting policy the implementation of regulatory practices will be problematic.
4. Ministry officials are rotated every two to three years to new divisions so that they become generalists. Their new positions can be entirely unrelated to their previous work assignments and they must once again adapt to new environments and to new managers, who are also generalists. Often, the officers receive a minimum of information from their predecessors about what transpired during their terms of duty. The new officers are expected to learn on the job and to acclimatize quickly. In the case of NISA, the supervision of nuclear facilities, nuclear

waste management disposal and the nuclear reprocessing plants could be inconsistent because of the continuous rotation of staff.

In all of the ministries, factionalism prevails and promotion is often predicated on the correct alignment with superiors who are members of the most influential faction. Officers come to understand that good standing among their superiors rather than the quality of their work is crucial to advancement up the ministerial pyramid. As is the case in all ministries, METI staff will comply with their superiors' directives even if they disagree in order to appear neutral and to prove loyalty.

In an attempt to garner more public support for the resumption of the reactors METI minister Kaida's announcement on 4 August that he was dismissing METI administrative vice-minister Kazuo Matsunaga, NISA director-general Nobuaki Terasaka and ANRE director-general Tetsuhiro Hosono for mishandling the nuclear crisis, was considered as merely cosmetic because the officials were set to retire at the end of the summer and it coincided with the annual reshuffle of personnel.

NISA: the tip of METI's nuclear tail

NISA is only a piece of METI's nuclear territory. METI and its nuclear energy-related organizations that are discussed in this chapter and in previous chapters have been integral to the nuclear industrial complex since the early 1950s. Since the research and development of nuclear energy is protected by the state as one of its industrial policies and since it is also connected to issues of national security, METI, as a key driver of nuclear energy, is intimately involved in the nuclear industry. Furthermore, its alliance with the electric power companies has been ongoing from the 1950s.

METI is the industrial ministry. Its bureaus administrate the manufacturing, energy, commerce and information industries, small and medium-size business development, the industrial science and technologies sector, economic and industrial policy planning, trade policy, and international trade and economic cooperation.

NISA's division is adjacent to METI's Agency for Natural Resources and Energy (ANRE) which has a staff of over 400 who work in four departments. The natural resources and fuel department includes divisions related to the procurement, distribution, mining and marketing of petroleum, natural gas, coal and minerals. There is an energy-efficiency and conservation division and a new and renewable energy division. ANRE's electricity and gas industry department comprises an electricity market division, a gas market division, an electricity infrastructure division (regional grids), a nuclear waste management office and nuclear energy policy planning division, and a nuclear fuel cycle industry division. ANRE controls Japan's energy sector and the development of energy resources.

METI's nuclear progeny: IAIs, industrial associations and research institutes

METI's sphere of influence is specific to the nuclear power industrial complex and its IAIs and industrial associations serve to expand the ministry's influence over the nuclear power industry. Moreover, METI's research institutes were the drivers of the development of nuclear fuel recycling and FBRs and the companies that are engaged in nuclear power often collaborate with the institutes, which are heavily funded by government.

The Japan Nuclear Energy Safety Organization[14] is a METI IAI that was established on 1 October 2003. With a staff of 426, including six senior managers, its activities are somewhat similar to NISA regarding the inspection and safety evaluation of nuclear facilities and emergency preparedness. According to the JNES website, its mission is to ensure nuclear safety in the use of nuclear energy and to analyze the safety of the designs of nuclear facilities and nuclear reactors. There is also a division that is engaged in the research of seismic safety which was set up after the Kashiwazaki-Kariwa accident in 2007.

The former MITI established the Radioactive Waste Management Corporation (RWMC)[15] in 1976 as an institute for the research of the treatment and disposal of radioactive waste. When the Radioactive Waste Final Disposal Act was passed in 2000 RWMC was designated for the administration of the funds for waste disposal. In 2005, when the Act for Spent Fuel Reprocessing Fund was enacted, RWMC was also designated for managing the reprocessing reserve fund.

In 2000, the former MITI established the Nuclear High-Level Waste Management Organization (NUMO)[16] for choosing deep geological storage sites for the spent fuel and for constructing and operating the facilities. Although applications for the selected sites were distributed from 2002 to thousands of towns none of the locations responded.

The Japan Atomic Industrial Forum Inc. (JAIF)[17] was founded as an independent non-profit organization in 1956 to serve as a platform for the peaceful use of atomic power. The 400 members include the heavy industries, electric machinery producers and large trading houses that are within METI's administrative jurisdiction.

It was explained in previous chapters that METI's regulation of the utility companies is strengthened through the relationship with FEPC and through the Electric Power Development Company. EPDC was renamed J-POWER[18] in 2005. J-POWER engages in the construction of nuclear power plants.

METI supports the Central Research Institute of Electric Power Industry (CRIEPI)[19] which operates a nuclear technology research laboratory and radiation safety center and a civil engineering laboratory which manages a nuclear-cycle back-end research center.

The end of the line: IAIs at the very heart of government

When it was revealed in April that former director-general of ANRE Toru Ishida had taken a job with TEPCO in January Yukio Edano announced at a news conference on 13 April that *amakudari* was 'socially unacceptable' and asked senior METI officials to temporarily refrain from taking jobs in power companies 'in order not to raise the public's suspicions'.

Edano questioned whether Ishida had entered TEPCO legally, pointing to the civil service regulation that requires retired ministry officials to migrate first to public corporations and wait for a period of two years before entering the private sector in order to loosen their ties with their former ministries. Ministry officials will try to bybass the regulation through methods that are described in Chapter 2. Edano later requested that ministry officials avoid employment in companies in the sectors that were within their ministries' administrative jurisdiction. However, since there is no law to prohibit this practice, bureaucrats normally find employment in firms with which they have had connections through their ministries and through their ministries' IAIs.

METI's nuclear energy-related IAIs, industrial associations and research institutes pervade the nuclear industrial complex and the constant contact between the heavy industries involved in nuclear power, electronics firms, the electric power companies and the ministry officials is inevitable. The relationship between METI and industries engaged in nuclear energy is further intensified when METI officials migrate from IAIs, industrial associations and research institutes to positions in the heavy industries or in the utility companies. The heavy industries that are engaged in the energy sectors have been collaborating with the Ministry of Industry since the Meiji Restoration and the relationship between METI officials and the industries is an abiding one.

As an example, Mitsubishi Heavy Industries engages in national and JNFL[20] projects in the construction of reprocessing plants (see p. 154).

It operates a nuclear fuel reprocessing plant in Tokai-mura for MOX fuel fabrication where the JAEA plant and the JCO reprocessing plant are also located. In 2006, Mitsubishi signed a contract with Kyushu Electric Power Co. to supply MOX fuel produced in its J-MOX plant for use in the Genkai pluthermal reactor, which was loaded in May 2009. It is a first for MOX to be used in a commercial reactor.

Norio Sasaki, the president of Toshiba Corp. which delivered the reactors to Daiichi, assured shareholders at a meeting in June, three months after the nuclear crisis, that the company would make permanent improvements in its reactors as required by the government's safety standards. He also said that when building plants in other countries, Toshiba would check the conditions of customers in each country. The company aimed to receive contracts to build thirty-nine reactors globally by FY 2015 amounting to ¥1 trillion ($12.43 billion).[21]

Traditionally, companies that hire retired senior officials from the ministry which administrates their industrial sector consider the employment as a part of corporate strategy to expand business opportunities. Companies that employ retired bureaucrats can facilitate the cooperation from the ministries to implement policies and regulations deemed favorable to business practices. The relationship may also assist the companies to receive government contracts, and in this case, perhaps subsidies for building nuclear power plants and fuel reprocessing installations. For the ministries, the *amakudari* system provides jobs and helps to facilitate cooperation from the firms to implement policies such as the formation of cartels, the rationalization of production or price-setting. The relationship can result in a mutual obligation/mutual protection/back-scratching mechanism that promotes strong vested interests in both camps.

Ministries do not like to lose control over any administrative territory. On 24 May, Kan's cabinet announced its plans to establish a special committee under the control of the Cabinet Secretariat to investigate the Daiichi disaster. METI officials were not pleased because the panel would be independent from the ministry and limit METI's and TEPCO's input. On 6 June, the Cabinet Secretariat's National Policy Unit, which is comprised of bureaucrats from the national ministries, submitted a paper entitled 'Regarding a Revolutionary Energy and Environment Strategy'. The paper recommended that a Council for the Realization of the New Growth Strategy be established and that the accident panel and the Japan Atomic Energy Commission be guided by an energy and environment committee within the Council. Although the paper designated METI minister Banri Kaeda and Masayuki Naoshima who served in both Hatoyama's and Kan's cabinets, and a former METI vice-minister to serve on the committee, Kan rejected the proposal. Instead, he called for the establishment of another committee to see whether NISA should be separated from METI, the assessment based partially on the results of the investigation by the accident panel. However, the panel was not established through legislation and therefore did not have a legal status. Since Kan was predicted to resign from office in August METI officials were not concerned about the temporary set-back.

Although Kaida pledged that national policy would not deviate from nuclear power as the mainstay of Japan's energy source, by June, the public and local government officials in some of the prefectures where the nuclear reactors were located were demanding a halt to operations at some of the nuclear reactors and the cancellation of existing plans for the construction of new reactors. Kan, pressing for the development of renewable energy sources, announced on 12 July that the government's current goal to increase Japan's reliance on nuclear power to 53 percent by 2030 should be dismissed in favor of alternative energy sources, which accounted for only 1.1 percent of Japan's power output. Kan conceded that Japan would have to depend more on fossil fuels in the interim.

Nevertheless, Kan's plans to shift national policy from nuclear to renewable were ambiguous and considered economically unsound by his economics minister Kaoru Yosano, not only because of the lack of specifics but also because the additional use of fossil fuels required by the utility companies would increase corporate taxes by 30 percent.

Since the public was opposed to the construction of new reactors and the future of MOX, FBRs, the operation of the Rokkasho J-MOX plant and the spent fuel storage facility were in the process of being reassessed, METI officials were eyeing greener pastures and the renewable energy industries, which were already in its portfolio.

METI, the creative ministry: the route to the renewable energy industrial sector

METI, as the energy ministry, had promoted renewable energy for years through its IAIs and research institutes which engaged in the development of alternative energy resources. METI officials saw a window of opportunity to create a new national policy targeting the problems of energy constraints through the intensive research and development of renewable energy technologies. On 2 June, METI announced plans to join with the private sector to develop renewable energy and power-saving technologies in order to ensure a stable and cost-effective power supply. METI's strategy called for cooperation between government, the private sector and universities. Out of the 2011 existing budget of ¥589 billion ($7.35 billion) for technological research only ¥30 billion ($375 million) was allotted to renewable energy, hence the funding would either be reallocated or provided through a supplementary budget.

The new strategy would also help to expand administrative jurisdiction by connecting the ministry to more companies in the renewable energy sector and to expand the budgets of its IAIs, research institutes and METI which was receiving sharp attacks from media and the public for its part in the nuclear disaster.

METI's announcement mentioned the Integrated Coal Gasification Combined Cycle (IGCC), a technology used for the production of a clean source of energy from coal. Coincidently, the Clean Coal Power R&D Co. Ltd. had been established in 2001 for the research and development of IGCC. The shareholders were the nine regional electric power companies and EPDC (now J-POWER) and supported by METI and the CRIEPI. Before its establishment it was known as the IGCC Association, which was established in 1986 and commissioned by the New Energy and Industrial Technology Development Center (NEDO) to perform tests in coal processing at a demonstration plant in Nakoso, Fukushima.

The research carried out at the Energy Technology Research Institute (ETRI) is pertinent to METI's strategy because it includes new energy technologies such as fuel cell systems, clean gas energy, thermal power and solar light conversion. ETRI is among the forty autonomous research institutes within the National Institute of Advanced Industrial Science and Technology (AIST).

The opportunity to increase the number of research facilities and perhaps more IAIs or subsidiaries of IAIs is not an unreasonable assumption. And even though NISA will be sectioned off from METI it is likely to be impossible to extricate METI and its officials from the nuclear energy sector. Cutting off a piece of the tail will only serve to regenerate it and to make the tail longer.

Don't blame the bureaucrats, blame the system!

Kan regarded the relationship between METI officials and the electric power companies as collusive. The relationship is the natural consequence of a system that had been set in the early 1950s and continued uninterrupted with the consistent support from law-makers who were also supported by the power companies and by the companies engaged in nuclear. METI's relationship with many of the companies involved in the nuclear industry can be traced to the Meiji period. The evolution of interpersonal networks between all of the players engendered a close-knit family, which is illustrative of the Japan Inc. model.

METI's IAIs and other public corporations are at the center of the nuclear establishment because they provide the ideal conditions for ministry officials to meet regularly with corporations involved in nuclear and to facilitate the migration of officials to positions in the electric power companies and business engaged in the nuclear industry thus expanding METI's influence over the nuclear energy sector. All of the ministries use their organizations in similar ways and can easily manipulate their activities to suit their needs and to justify budgets.

The *amakudari* system offered university graduates the incentive to enter the ministries where working conditions can be demanding. The institutional structure of both Japanese private corporations and government agencies is hierarchical and promotion is based on the seniority system, making advancement slow. By the time a career official reaches fifty-five he or she can expect retirement within a few years to make way for a junior official. Work days can end at midnight and there is usually no overtime pay. Administrative support is random until the officials reach a relatively high rank. Wages are also lower than in the private sector. But the anticipation of achieving influential positions in government and lucrative post-retirement positions give them the incentive to adapt. Their superiors look after their welfare, perhaps arranging a two-year respite at a university overseas or at one of the branches of an IAI or research institute, a good number of them located in foreign countries.

Retired officials can expect a comfortable job waiting for them in the private sector or in a public corporation with better salaries and far easier working hours. The system was established by the civil service system and supported by obliging law-makers and the electorate until the 1990s when scandals concerning the collusion between ministry officials and industry began to tarnish the image of elite bureaucrats whose ethics had previously been considered spotless. Nevertheless, by the 1990s, it was too late to reform *amakudari* because it had become an indelible part of the model of administration.

Politicians have tried to loosen the bureaucracy's control over policy-making by banning *amakudari* and dismantling IAIs. However, it is absolutely essential to provide alternative incentives in order to promote careers in government to talented graduates who now prefer to enter the private sector for more satisfying careers. It is difficult to conceive of anything better (or more comfortable) than *amakudari*.

8
The Japan System: Indestructible but Destructive

The events of 11 March 2011 are considered to be Japan's worst disaster since the Second World War. The earthquake–tsunami and the nuclear crisis impacted severely on a deflated economy that carried a sovereign debt equaling twice the annual GDP. A review of the problems that plagued Japan's economy since 1990 and the issues that frustrated the implementation of fiscal and monetary policies that would put the economy back on track will enable a reasonable assessment regarding the pace of Japan's post-11 March economic recovery.

Prior to the bursting of the real estate asset-inflated bubble in 1989 Japan was not burdened with sovereign debt. By the 1970s, the Japanese were experiencing a rapidly expanding economy and disposable income and enjoying the reputation for producing for world markets the finest cars and electronics and high-tech products. Their business model seemed a perfect combination of elite civil servants and big businesses driving the economy forward while a single political party gave unwavering support to policies that were planned and implemented by the bureaucracy. The majority of the post-war electorate was satisfied with their Japan Inc. model.

The 1985 Plaza Accord and the floating of the dollar triggered what is referred to as Japan's 'asset-inflated bubble economy'. Corporate overexpansion in Japan during the bubble years of 1986–90 was funded by substantial loans from national and regional banks. Many of the loans were underwritten by the rising value of real estate. In the 1980s the price of land began to spiral and this trend was accelerated by fiscal policy implemented by the Bank of Japan's policy of loosening interest rates. In 1990, the BOJ reined in interest rates and the MOF introduced restraints on bank lending to real estate-related industries: real estate agencies, construction companies and the non-banking subsidiaries of banks. After the rates were tightened the bubble burst and real estate values fell heavily and the banks were left with a substantial body of non-performing loans or 'bubble loans'.

Japan's first 'lost decade'

In May 1991, the BOJ issued a warning that the price of real estate was excessive and that if it was acknowledged that the rise in real estate values was not sustainable the consequences would be as severe as was the case in the secondary banking crisis in the UK in the 1970s. However, BOJ and MOF officials were not prepared for the sudden sharp decline in real estate prices and the effect on financial stability.

The Japanese government claimed that through the 1990s it injected ¥9,000 billion ($83.5 billion) into banks in order to avoid a banking crisis.[1] Although a substantial amount, this was a very small proportion, and considerably less than 10 percent, even on government estimates, of the total of non-performing loans held by the banks. The Japan Economic Institute, an organization under the auspices of the Ministry of Foreign Affairs, reported in August 2000 that the Financial Services Agency declared the amount of problem loans for all banks as of March 2000 equaled ¥81.8 trillion ($743.6 billion) compared with ¥80.6 trillion in 1998.[2] In his address to the Japan–US Business Conference on 17 February 2002 the chairman of Keidanren, Takashi Imai, claimed that the banks had disposed of ¥75 trillion.

Due to conflicting reports it is difficult to calculate the actual amount of NPL. Lincoln (2003) stated that the government estimate of NPL for the end of September 2002 was ¥40 trillion ($330 billion) and that during the decade 1993–2003 the banks had written off ¥83 trillion (about $700 billion) of NPL. However, he contended that there was speculation in the private sector that the 'real amount' of these disposed NPL could be double or triple the official report.

Even the large amount of loans that has been written off does not necessarily represent loans that have been resolved. In many cases, the banks have simply set aside reserves equal to the amount of the loan while taking no action against the borrower to recover collateral, leaving 'zombie' borrowers – effectively bankrupt but still in business.[3]

Kobayashi (2004) maintained that during the 1990s the banks simply denied the bad-loan problem and that, although the official declaration of non-performing loans in 2004 was ¥35 trillion, the private sector put the figure at ¥87 trillion.[4]

Until the 1990s, Japanese banks understood that they could rely upon the BOJ as their lender of last resort and that both the BOJ and MOF regulators were their protectors who shielded them from close scrutiny by outside auditors. Many retired bureaucrats from MOF and retired BOJ executives were employed in top management positions in metropolitan banks and financial institutions. As of 1992, there were seventy-eight former MOF officials and sixty-four BOJ officers on the boards of 115 listed banks: 51 percent of the banks were employing former MOF officials on their boards, 44 percent had former BOJ executives on their boards, and 69 percent had an official from either the MOF or the BOJ. The consequence of the *amakudari* was poor screening practices.[5]

The first 'lost decade': institutional paralysis

The 1990s are referred to as Japan's 'lost decade' because the struggle for power between factions within the major political parties as well as the vested interests among the members of Japan Inc. served to continuously interfere with the implementation of fiscal policies that would bring Japan out of deflation and ignite an economic recovery.

In 1990, the BOJ tightened monetary policy but when the recession hit in 1993 the overnight call rates were lowered and the BOJ purchased more JGB to provide support for the growing government debt. Japan's deflationary cycle began in 1995 and continued for the next fifteen years. In March 1999, the BOJ lowered its uncollateralized overnight call rate to 0.1 percent and began its expansionary monetary policy tool, quantitative easing (QE), in a desperate effort to stimulate Japan's deflated economy by purchasing government bonds.[6] However, there was no political will in the financial system to reform practices because of the banks' rigid and inward corporate culture that had been developing since the 1950s and because of the profound vested interests between the lenders and the borrowers through cross-shareholdings. The government's fiscal policy did not go beyond the release of countless stimulus packages for public works projects in order to keep unemployment figures low.

The BOJ's soft monetary policy and the government's fiscal stimulus packages in the 1990s failed to lead Japan out of deflation and promote stable growth. Again, between 2001 and 2005 the central bank executed another round of QE and pumped ¥40 trillion ($493.6 billion) into the financial system by purchasing JGB at 0.1 interest rates in order to induce commercial banks to increase corporate lending. It also purchased commercial paper to create further liquidity. Nevertheless, QE did little to alleviate deflation because the banks used the money to write off their load of non-performing loans. Toshihiko Fukui continued to support QE policies when he became director-general of the BOJ in 2003.

Since the mid-1990s, analysts have written prolifically about the reasons for the rupture of Japan's asset-inflated 'bubble' economy. But in 1988, the international business community, preferring to focus on Japan's dynamic economy and trade surplus, ignored the looming rupture of the real estate and stock markets, Even though Japan was the world's largest creditor, the economy was clearly overheated. The Japanese themselves called their economy a 'bicycle economy' (*jitensha keizai*) because they had to pedal continuously to keep it running at high speed.

Although Japan was in recession and beset with grave economic problems, analysts remained very positive about future growth. Japan's economic problems were attributed to macro-economic reasons emanating from inflated property and stock prices. The possibility that major structural problems in the Japanese governing system itself were contributing factors was generally not considered until fiscal stimulus packages not only failed to ignite the economy but also sent government debt sky-rocketing. There was also little recognition that the network of bureaucrats throughout business and the political community could act to ossify the political economic system and was a major factor in Japan's inability to take the necessary measures to salvage the economy.

Japan's second lost decade: institutional paralysis

Japan's second lost decade revealed that the ties between the bureaucracy, the business community and elected officials served to sustain a rigid and inward model of government administration which paralyzed the decision-making process and prevented the implementation of the reform of the system to accommodate the ever-changing demands on the domestic front and the external impacts from international markets.

Koizumi's campaign mantra 'structural reform with no sacred cow' and 'no growth without reform' promised fiscal and state sector reforms and the tackling of the non-performing loan problem. Koizumi was a flamboyant politician and very popular among voters. However, when he entered office in 2001 Japan's economy was in recession. The annual GDP growth rate had contracted by 0.5 percent (the first annual drop in three years), and the sovereign debt had climbed to 130 percent of the GDP. Manufacturing output plummeted by 1.5 percent from its mid-2000 peak – a thirteen-year low. Property prices fell by 5.9 percent, the fastest decline in nine years, with commercial property falling by 8.3 percent, which was 63 percent below the peak value registered in 1993.

In 2002, as the public debt climbed to 140 percent, Japan's continued deflation took its toll on the real estate market. The GDP growth rate was 0.1 percent. The 2002 Assessment and Recommendations for Japan's economic recovery released by the OECD in October 2002 projected that Japan's economy would grow only by 0.5–1.0 percent per annum until 2004, and that because of the general slowdown of the global economy, Japan's deflation would worsen. It urged the government to take radical steps to bring about the end of deflation. The report also stated that the economy could be revitalized if structural reforms were implemented but the OECD also admitted that reforming the system would be difficult because of political opposition and vested interests. It warned that time was of the essence.

Since its monetary policy instruments had been ineffective in ending deflation and sparking growth, in May 2003 the BOJ took rather extreme measures at the request of the MOF and intervened in the foreign exchange market, selling a massive ¥20.2 trillion ($177 billion) in order to depreciate the yen further against the dollar, thereby making Japanese exports more competitive in foreign markets. The average yen–dollar exchange rate in 2003 was 115.82. The BOJ continued its overnight call rate at near zero percent, a policy that it implemented in early 1999.

Although in 2003 the GDP growth contracted 0.3 percent, Japan's economy began to pick up during the next three years. During the first three months in 2004, the BOJ purchased ¥14.8 trillion ($139 billion) but the intervention did not help to weaken the yen to the desired levels. In 2004, the average yen per dollar exchange rate was ¥107.49. In September 2004, the BOJ raised the overnight call rate to 0.5 percent. In 2005, the yen weakened to ¥110.85 per dollar. Deflation continued to eat away at growth.

In 2004, the GDP growth rate bounced back by 2.7 percent and in 2005 by 2.9 percent due to fiscal stimulus packages and secondary budget top-ups, foreign exchange intervention and soft monetary policies. In 2006, Japan's GDP growth decreased by 2.6 percent and by 2.2 percent in 2007. The following financial year saw deflation continuing and the sovereign debt climbing. The government had hoped that deflation, the weakening yen and the fact that Japan's trade surplus was steadily decreasing would encourage manufacturers to export more, a strategy that had been fundamental to Japan's economic policy during the post-war period. However, the dependence on external demand to revitalize the economy would be only a temporary solution. The real problems were structural and the solutions hinged on the implementation of structural reforms.

Although Koizumi's reforms focused on the public sector, outstanding government debt had climbed to 150 percent of the annual GDP. Koizumi was engaged in a tug-of-war with the ministries to dismantle Special Corporations and to cut public funding to public corporations but the bureaucracy remained in charge of the reforms. Koizumi vowed to cut public spending and he did. But the consequence was an asymmetric economy. Trade with China gave Japan's economy a push but only the larger corporations benefited while the smaller companies languished. Koizumi's privatization of Japan Post and the downsizing of FILP also weakened the LDP at the polls in local elections.

Koizumi's legacy to his successor Shinzo Abe in 2006 was an economy in deflation and a 150 percent sovereign debt because of the continuous release of fiscal stimulus packages and supplementary budgets to subsidize prefectural economies. Abe's administration was wracked with scandal leading to the LDP's humiliating defeat at the polls and to Abe's resignation. When Takeo Fukuda entered office in September 2007 the voters were already disenchanted with the LDP and political gridlock. The Japanese who were near retirement age were particularly displeased with the new health-care program that required premium payments to be deducted from pension benefits.

From the end of 2007, the government began releasing monthly data that forecast a bleak year for Japan's economic growth. In February 2008, the Cabinet Office announced that consumer confidence had declined to its lowest level since 2001 at 33.9 and was expected to deteriorate further. The Cabinet Office announced in April that economic growth had 'paused' and that business confidence was weak due to fears of a recession in the United States because of the subprime mortgage crisis and the steep price increase of raw materials and petroleum. As the yen strengthened to ¥102.80 per dollar, the BOJ also predicted much slower growth. Prime Minister Fukuda suffered the same fate as Abe and resigned after only one year in office. Taro Aso took the helm in September 2008.

The IMF urged the BOJ to further lower overnight call rates. On 19 December, the BOJ lowered the interest rate to near zero percent three days after the US Federal Reserve lowered its policy rate to the same level. The BOJ also purchased commercial paper to help companies fund daily operations and increased the purchase of JGB to continue its QE policy.

Data released that month showed that steel production fell 12.9 percent in November, the steepest decline in ten years. The BOJ predicted severe economic conditions in 2009 and 2010 but successive administrations' attempts to implement sound fiscal policies were inhibited by the usual political bickering and vested interests.

The DPJ's landslide victory in the Lower House election in August 2009, ending the LDP's domination of the National Diet for over half a century, was predicated on the DPJ's promises in its manifesto to revitalize Japan's economy through reducing the power of the bureaucracy, ending political corruption and increasing financial aid to families. There were great expectations of a new era of institutional and economic reforms that would help to encourage sound fiscal policies. But the mechanisms in Japan's governing system that had been stymieing the implementation of fiscal policies and reforms prevailed. In 2009, as the American economy slumped and the dollar weakened, the yen continued to strengthen against the dollar (¥93.50 per dollar). The currency appreciated further to an average ¥87.10 per dollar in 2010, which was a low for the currency because it had appreciated to ¥83 per dollar in September.

After a six-year respite, on 15 September 2010 following Kan's entrance as prime minister, the BOJ sold $1 billion worth of yen in three successive sessions of foreign exchange intervention. The yen depreciated 3 percent against the dollar. Nonetheless, it was predicted that the intervention would not be effective to make a difference to its future value. The average exchange rate in 2010 was ¥87.10 per dollar. In the last quarter of 2010 the GDP contracted by 0.3 percent. The annualized contraction was 1.1 percent due to weak consumption.

During the first three months of 2011, the exchange rate hovered at ¥81 per dollar, testimony to the fact that the BOJ's foreign exchange intervention was proving ineffective. Exporters were pressuring government to take additional action to weaken the currency.

As was stated in Chapter 6 the BOJ on 15 February 2011 upgraded its general assessment of the macro-economy, indicating that the country was gradually emerging from recession. However, the 'emergence' relied heavily on the increased demand from emerging markets for goods from the larger exporters and not on domestic consumption. Indeed, about 90 percent of the 1,333 companies listed on the First Section of the Tokyo Stock Exchange (TSE) reported that collectively they had made a profit of ¥14.78 billion ($184.76 million) in FY 2010, which was a vast improvement from FY 2009.[7]

Earthquake–tsunami and nuclear crisis: impact on industrial production

From February to the end of April factory production was the lowest since the end of the Allied Occupation in 1951. On 30 April as power shortages throughout Honshu and the nuclear crisis continued, government reports showed that factory production had plunged a record 15.3 percent from February and household consumption had dropped 8.5 percent from the same period a year earlier. Retail sales fell at the fastest rate in thirteen years. The bad news continued throughout June with a 30 percent decrease in exports, which account for 15 percent of Japan's GDP.

Directly following the earthquake, the yen began to appreciate and on 17 March, when it hit ¥76.35 per dollar the BOJ injected ¥7 trillion ($83.33 billion) into the money supply, purchasing dollar/yen in an effort to stabilize financial markets and calm investors. The United States, Britain, Canada and the European Central Bank agreed to join Japan in the intervention. The yen weakened 0.3 percent.

On 11 April, the BOJ, while calling for additional fiscal stimulus packages, downgraded its economic assessment for seven out of Japan's nine regions. Tokai, the center of automotive production, reported weak output mainly due to the damage to facilities, supply-chain disruption and limitations on the use of electricity. Companies, particularly manufacturers of cars, semiconductors, electronic components, optical products and paper, experienced huge difficulties restarting operations which impacted on supplies of parts to their domestic and foreign plants.

The BOJ anticipated in April that once production was restored, the economy would begin to recover but manufacturers such as Toyota, Honda, Sony, Toshiba and Hitachi with plants located in the northeast had been forced to limit production mainly because of power disruptions. A shortage of parts affected both domestic producers throughout Japan and the exporters' customers. The parts shortages delayed the recovery of production throughout Japan and led to Japan's largest trade deficit on record in April and May. South Korean parts manufacturers were already receiving orders from the large Japanese producers by May and there was deep concern among auto subcomponent manufacturers whose installations and equipment were damaged by the earthquake that they would be hard-pressed to recover if the big producers outsourced parts from other countries or shifted production overseas.

The automobile industry accounts for about 8 percent of the domestic workforce. Toyota, the world's largest car producer, was forced to cut production by 50 percent in Japan and China and 70 percent in the United States until 3 June but extended the period until July. The company expected to normalize global production by November–December but by June anticipated the return to normal production in the United States. Honda discontinued production of cars until May at plants in the United Kingdom and in the United States also, reflecting the disruption in the supply chain. Domestic automobile sales fell more than 50 percent in March–May. In April, Toyota sales fell by 69 percent, Honda sales dropped 47 percent and Nissan sales declined 37 percent. However, the plunge in sales was not entirely due to the earthquake because it marked the eighth consecutive month in the decrease of sales. Japanese exports dropped by 2.2 percent in March from a year earlier.

Toyota reported in May that the cost related to the earthquake–tsunami was estimated at ¥100 billion ($1.23 billion). Nonetheless, Toyota's profits had already plummeted 77 percent from January–March due, in part, to the strong yen, recalls of automobiles in the United States, slow domestic sales and the end of government tax-breaks for the purchase of environmentally friendly vehicles. In June, the Automobile Dealers Association announced that domestic car sales continued to fall for the tenth straight month.

Although the parts shortage disrupted global operations and impacted negatively on sales in overseas markets, American auto manufacturers took advantage of the Japanese producers' forced respite. The earthquake had no major impact on production lines at Ford Motor Co. or Chrysler Group LLC. While Toyota's sales declined compared to the same period a year earlier, General Motors Co. saw global sales expand 11 percent during the first quarter. According to car dealers in the United States, although the earthquake impacted on Japanese car sales, the Japanese manufacturers were already losing their market shares because of Toyota's recalls in 2010 and because Honda's designs were no longer as popular. Customers were trading in their Japanese models for cost-efficient foreign models. Toyota and Honda will struggle to recover American market shares from Hyundai and Kia, South Korea's number one and number two car manufacturers respectively. Also, the currency's appreciation, which marked a record post-war high against the dollar at ¥75.95 on 19 August will affect sales in global markets[8] and erode profits.

Five major electronics manufacturers posted losses of ¥354.5 billion ($4.4 billion) in FY 2010 because of the earthquake. The consolidated financial reports showed operating profits of Hitachi Ltd., Panasonic Corp., Toshiba Corp., Fujitsu Ltd., NEC Corp. and Sharp Corp. fell more than ¥100 billion ($1.23 billion). Companies postponed the launch of new appliances due to the shortage of parts.

The disruption in the supply chain impacted on foreign clients as well, particularly in the electronics and automobile sectors. Sony Ericsson missed its shipment forecast because of difficulties sourcing display, batteries and other parts. On 16 July, the company declared a $71 million loss until June. Apple Inc. announced that margins would be under pressure because of the shortages of electronic gadgets. Motorcycle manufacturer Harley-Davidson lowered shipment forecasts due to problems procuring radio components. General Motors Co. was concerned about parts shortages and temporarily shut down operations in one of its plants in Louisiana. The semiconductor industry was significantly affected because production requires precise control of electric currents and machine tools. Texas Instruments experienced disruption in production of two chip-manufacturing plants in Japan that supplied 10 percent of the company's total output. Both Japanese and foreign companies, which had procured mainly from Japanese suppliers, began sourcing parts from other suppliers in Asia.

The Cabinet Office's Economy Watcher Survey which measures the assessment of current economic trends by retail store owners, taxi drivers and people living in economically active areas was conducted for end of March. It revealed that business confidence in Tohoku plunged from 47.2 pre-earthquake to 26.6 post-earthquake not only because of the damage which the earthquake and tsunami caused to production facilities but consumption had deteriorated as well due to continuing concerns about the state of the economy.

Besides Tohoku, the tourist industry throughout Japan experienced an immediate impact. In March, there were 390,000 cancellations in Tohoku and Kanto districts and 170,000 in the rest of Japan, a 73 percent fall from the same period in 2010. Hiroshi Mizohata, commissioner of the Japan Tourism Agency, reported that the earthquakes and the concern over the continuing dispersion of radioactive materials were severely affecting tourism leading to a sharp drop in demand among Japanese and foreign tourists. If the situation did not improve small hotels and inns would experience big losses in revenue. The situation improved slightly in April with a drop of 62.5 percent but it was the largest year-on-year drop on record. Foreign visitors to Japan in May fell 50 percent, with South Korean visitors in the lead followed by the Taiwanese and the Chinese who are significant to Japan's tourist trade. There was little improvement throughout the summer due to the expensive yen and the concerns about radiation contamination.

Confidence in western Japan, which was not directly affected by the disaster, also plunged because of the rolling blackouts. Disruption in supplies of parts produced in Kanto-Tohoku to producers in western Japan had forced a shutdown of some operations. It was projected that the manufacturing sector would be affected by power shortages for at least five years, impacting on production and on the supply chain.

In April, radioactive contamination was detected on non-food products. The Chinese government and importers in other Asian countries requested certificates from exporters ensuring that the goods were free of radiation contamination. The Japan Automobile Manufacturers Association (JAMA) announced on 18 April that automobile manufacturers were testing their domestically produced vehicles for radiation before shipping them to domestic and overseas markets in order to alleviate the concerns of customers about products made in Japan. However, on 1 May customs agents at the port of Iquique, Chile, detected low levels of radiation in automobiles that had been transported from the Port of Yokohama by a South Korean freighter. Although the cars were hosed down on board the ship so that the radioactivity would be contained in the vessel, port workers protested that their health was at risk. The episode was the first time that radioactivity was discovered in Japanese vehicles since the outset of the nuclear crisis.

The BOJ reported on 12 May that lending fell to 1 percent from the previous year to ¥395.65 trillion ($4.93 trillion) and money supply grew by 2.7 percent. Commercial loans fell for the seventeenth straight month. At the end of June, the Japanese Bankers Association reported that the outstanding balance at 119 banks was ¥415.14 trillion ($5.109 trillion), a fall of 0.8 percent on the year because of low corporate demand for capital for the twentieth straight month. However, the balance of deposits rose for the fifty-seventh straight month at a rate of 2.4 percent or ¥582 trillion ($7.25 trillion).[9] The rise reflected the high degree of uncertainty among businesses and households regarding Japan's economic recovery.

Although the costs for reconstruction of infrastructure, housing and installations can be estimated, it is difficult to predict how job losses and industrial devastation in the afflicted areas will affect local residents. Nomura Research Institute, which is associated with Nomura Holdings Inc., projected in early April that the number of job losses in the Tohoku coastal area would rise above 81,500. However, the number did not include jobs in the farming and fishing industries, industries that were thriving before the earthquake. Nomura found that jobs in the manufacturing, wholesale and retail sectors would be the hardest hit and urged government to implement policies to create jobs, including in the forestry, farming and fishery sectors.

Nomura Holdings is indicative of Japanese brokerage houses which earn most of their revenue domestically.[10] The drastic cuts in production by domestic companies led to a contraction in demand for underwriting and takeover consulting services. While suffering losses domestically, Nomura was also experiencing competition in overseas markets. The company's purchase of Lehman Brothers' European and Asian businesses in 2008 failed to expand operations as had been forecast. Since the purchase of the bankrupt company, shares of Nomura dropped 37 percent during 2009–10 and continued falling so that by end of FY 2010 Nomura's shares had fallen 71 percent. 176 other domestic brokerage firms or 61 percent suffered pre-tax losses of 74 percent from the previous year.

Impact on sovereign debt: out-of-sight, out-of-mind

In March, the national debt, including government bonds, loans and financing bills, climbed to a record high of ¥924.35 trillion ($11.45 trillion), which was ¥5.2 trillion ($64.4 billion) higher than at the end of December 2010.

The government calculated the cost of reconstruction of social infrastructure, corporate facilities and housing in the affected prefectures at ¥16.98 trillion ($210 billion). In May, the Diet passed an extra budget for FY 2011 of ¥4.02 trillion ($50 billion) to support the initial reconstruction efforts. The second extra budget of ¥2 trillion ($24.9 billion) was expected to pass in June but was delayed until 26 July. The third extra budget proposed by the DPJ in June was ¥10 trillion ($124.61 billion)[11] the largest recovery package. The government, perhaps in a state of denial about the consequences of public spending during the last two decades, contended that the Japanese would be able to assume the burden through the issuance of government reconstruction bonds and the levying of a temporary reconstruction tax. Many politicians were opposed to the reconstruction tax.

Both the IMF and the OECD urged a 15 percent sales tax because government would have to reallocate expenditure and increase revenue to fund the reconstruction efforts. Both organizations expressed concerns that Japan, with a public debt of over 200 percent of the annual GDP, the highest among the industrialized nations, might not be able to bear the enormous costs of the reconstruction.

On 27 January, the credit rating company Standard & Poor's downgraded Japan's long-term sovereign credit rating for the first time since 2002 from AA to AA−. S&P assessed Japan's fiscal deficit would remain high and weaken the 'government's already weak fiscal flexibility'.[12]

On 22 February, Moody's Investment Service downgraded the outlook on Japan's Aa2 rating from 'stable' to 'negative'.[13] The reasons given for the first negative action since 2002 was that there was no indication that the Japanese government would implement the tax reforms needed to cut the sovereign debt, the average interest rate on the outstanding debt was only 1.5 percent, and interest payments were 25 percent of national tax revenue.

On 21 April, Standard & Poor's also downgraded its forecast for Japan's sovereign debt from 'stable' to 'negative' because of the reconstruction costs and because the government was having difficulties consolidating fiscal policies to cut the debt prior to the earthquake:

> The downgrade reflects our appraisal that Japan's government debt ratios – already among the highest for rated sovereigns – will continue to rise further than we envisioned before the global economic recession hit the country and will peak only in the mid-2020s.

On 19 May, the Japanese government announced that the economy had contracted for two consecutive quarters and had slid back into recession. The GDP shrank 0.9 percent in the first quarter, realizing an annualized contraction of 3.7 percent. The decline in consumption was 0.6 percent (private consumption accounts for almost 60 percent of the economy). The trade surplus shrank by 34.3 percent in March from the same period a year earlier. Describing the economic conditions as 'severe' it cautioned that recovery could be further delayed due to the temporary immobilization of nuclear power plants which were undergoing safety inspections.

On 28 May, directly after Fitch Ratings lowered its forecast of Japan's sovereign debt from stable to negative, the BOJ, referring to the economic risks related to the nuclear crisis and the sovereign debt, reported Japan's fiscal condition as 'very serious'. The governor Masaki Shirakawa expressed concerns that Japan was losing market confidence and if the government failed to deal with its finances, the value of government bonds (JGB) would fall. Shirakawa was also worried that the weakening of confidence in private-sector financial firms, which were heavily invested in JGB, would lead to higher borrowing costs. The economy would suffer, tax revenue would fall and the government would have difficulty paying off its debt. In July, the BOJ was not prepared to guarantee more JGB to fund growth because it would undermine the trust in the currency, increase long-term interest rates and pose difficulties selling JGB on the market.

On 31 May, Moody's Investor Service, citing the slower-than-forecast pace of production, higher unemployment figures and an anticipated national debt for FY 2011 of 218 percent, reviewed the possibility of downgrading Japan's debt rating. Moody's expressed concerns that the expenditure for the reconstruction, which included taking on part of TEPCO's compensation payments, would inhibit Japan's future growth and that the ongoing political bickering and power struggles between the political parties in the Diet would impede decisive action regarding fiscal policies.

On 9 June, Standard & Poor's announced that it was considering lowering the long-term sovereign credit rating again because the debt would be higher than it had anticipated in January and because of the global economic downturn. Since prices in Japan had been stagnant during the last decade, it was unlikely that inflation would eat away at the debt. The rate of medium-term growth was predicted to be 1 percent because of the rise in the numbers of pensioners (3 percent in 2010), and the decrease in the workforce.

On 27 June, Moody's voiced concerns about the sovereign debt and announced that it was considering lowering Japan's long-term credit rating again, claiming that Kan's refusal to resign was delaying action by government to endorse plans to reduce the public debt and to make strong policies that should include the reform of social welfare spending and the bolstering of tax revenue. It pointed to indications that Japan was entering a third 'lost decade'.

On 24 August, Moody's downgraded Japan's credit rating from Aa2 to Aa3, which was the same level as the S&P's credit rating for China. It also cut its credit ratings on Bank of Tokyo–Mitsubishi UFJ and Sumitomo Mitsui Banking Corp. from Aa2 to Aa3 and both Mizuo Bank and Mizuho Corporate Bank from Aa3 to A1.

Although Keidanren was also promoting an increase of the consumption tax, members of both Japan's ruling party the DPJ and the LDP were opposed mainly because it was politically unpopular. The central bank was opposed to guaranteeing the debt because it would trigger inflation and emphasized the need to raise taxes to help fund industrial recovery. However, the growing anxieties about the heavy burden of financing the costs for the reconstruction, the escalating welfare costs for the elderly, and the downgrading of Japan's sovereign debt by key credit rating companies compelled the government to support a tax increase, a policy tool that the IMF and the World Bank had been urging since the end of the first lost decade.

Political bickering: impact on rapid recovery

In May, members of the LDP and the New Komeito were committed to forming a grand coalition with the DPJ and an extension of the Diet session, which was scheduled to end on 22 June, to deal with reconstruction bills but only on the condition of Kan's resignation. Despite the urging from DPJ colleagues, Kan refused to budge unless a secondary budget, a special bill to issue deficit covering bonds to cover at least 40 percent of the FY 2011 third budget and a renewable energy bill were passed. The LDP and New Komeito refused to negotiate and the usual political bickering and gridlock continued. On 22 June, Kan agreed to resign within seventy days in August. However, a week later, in keeping with his proposal that Japan should become a nuclear-free society and shift to renewable energy, he demanded that all of the reactors be put through a series of stress tests which would be assessed by government before the power utilities would be permitted to resume operations. This unilateral action was regarded as not only entirely unrealistic because Japan's renewable energy sector provided only 1.1 percent of Japan's power output, but also resulted in increasing opposition to his administration.

The Diet session was extended but the LDP and the New Komeito voted against the continuation of Kan's administration. Although a disaster relief budget of ¥2 trillion ($25.46 billion) was passed on 25 July to help Tohoku earthquake victims, political squabbling continued to impinge on the passage of the third supplementary budget of ¥10 trillion, the largest of the three supplementary budgets for FY 2011 and fundamental to the support of the quake reconstruction package. Ministry officials ceased visiting Kan's office because he was regarded as a lame duck. The resignation by Ryu Matsumoto, Kan's newly appointed reconstruction minister after only one week on the job did not help Kan's popularity either. His first visit to Miyagi Prefecture to meet with Miyagi governor Yoshihiro Murai was televised. Viewers were not impressed by his imperious attitude toward the governor coupled with his warning that the government would not support Tohoku unless the local leaders came up with good ideas about post-disaster reconstruction.

The National Policy Unit that was set up in 2009 under the control of the prime minister was losing the majority of its experts who had been recruited from industries to advise the cabinet because their companies wanted to distance themselves from Kan. By July, the organization was in disarray.

On 22 July, in order to bring the LDP on board to pass the bill for the issuance of deficit-covering government bonds and the measure for the development of renewable energy, the DPJ formally acquiesced to LDP demands and withdrew from its 2009 DPJ manifesto the government-funded child welfare allowance and other key pledges, which was a major concession to the party that had ruled Japan for over half a century. On 25 July, the Diet passed a ¥2 trillion second extra budget, one of the conditions that Kan insisted upon before leaving office. On 9 August the

LDP and New Komeito agreed to consider the third ¥10 trillion budget to be covered by deficit-covering bonds and supported with a temporary reconstruction tax hike on the condition that the DPJ revise its stance on other pledges made in its 2009 manifesto such as the discontinuation of highway tolls. The members of all parties hoped that the highly unpopular Kan would resign as soon as the bill passed both houses.

The desperate conditions of the evacuees stranded in evacuation centers in Tohoku and the plight of Fukushima residents who were waiting for government aid and compensation payments remained secondary to political power plays.

Big business was also demanding Kan's resignation. Hiromasa Yonekura, the chairman of Keidanren, criticized Kan for not standing down in June and called his plans to force the utility companies to buy renewable energy generated by businesses and households irresponsible. Yonekura claimed that the plan would lead to higher energy prices and increase the financial burden on companies forcing them to move overseas.

The BOJ governor Shirakawa told reporters on 12 July that the shutting down of the power plants would lower Japan's long-term growth prospects. He also expressed concerns about the impact of the yen's appreciation on exports. The government's monthly economic report released in July remained unchanged from June citing the continuing shortage of electric power supply and appreciation of the yen to an average high of ¥78 per dollar as having a negative impact on Japan's recovery. Its July report conceded that the quake and nuclear crisis would put downward pressure on medium to long-term economic growth.

Too little too late?

In 2009, METI formed an Industrial Competitiveness Committee under its Industrial Structure Council for the preparation of its Industrial Structure Vision, released in June 2010. The vision stated METI's objectives to 'launch a nation-wide effort to boost its industry's global competitiveness' to stimulate Japan's economic recovery. METI's aim included a shift in industrial structure to encourage the 'potential strengths in businesses', the creation of jobs 'by aggressive globalization and by building world-class business infrastructure' and to change the government's role to 'win the fierce competition to acquire added value among countries'.[14]

The vision, which described Japan as a 'deadlocked economy' included data from the IMF World Economic Outlook Database that showed the general deterioration of Japan's competitiveness in global markets and the effects on the stagnated economy. IMF figures revealed that Japan's global ranking in terms of GDP per capita was third in 2000, falling to twenty-third in 2008. In 1990, Japan's share of the global GDP was 14.3 percent while in 2008 its share had decreased to 8.9 percent. The vision reported that, although the economy was growing during 2002–7, wages remained stagnant or had declined. The number of companies going out of business or closing down exceeded the number of startups. MOF data revealed that while outward investment had increased, domestic capital expenditure had decreased. From 2006, domestic investment plummeted 37 percent. There was a significant shift by businesses to produce abroad.

Key manufacturers had been transferring operations overseas for a number of years to take advantage of cheaper labor, the availability of natural resources and to accommodate the currency's appreciation. In 2011, Toyota was producing 50 percent of cars outside of Japan and Honda and Nissan were each producing 70 percent of their vehicles abroad. Power shortages are predicted to continue for the next five years and the possibility of more disruptions in the supply chain will pressure companies to move their installations abroad, resulting in a further hollowing-out of Japan.

In June 2011, METI announced plans to release a package of subsidies to pay firms to remain at home. It was hoped that the subsidies would encourage producers to start new operations in regions throughout Japan. The new measure was included in the 'Industrial Structure Vision'. However, the package may come too late since producers were already setting up facilities elsewhere.

METI's 2009 vision also called for increasing foreign investment through the creation of an environment conducive to foreign businesses operating in Japan such as offering corporate tax cuts and building international airports and harbors. However, inward investment, which was 3.6 percent in FY 2010, froze as foreign investors began to reconsider post-earthquake investment. Japan had already lost its position as Asia's financial hub to Singapore and Hong Kong and foreign multinationals were relocating more of their divisions out of Japan after the earthquake hit.

The yen's appreciation and turbulent exchange rates were hampering exporters' competitive edge, encroaching on market shares in foreign markets and eating into profits. For example, Toyota based FOREX on ¥80 per dollar. The company estimated that during FY 2011 a one yen gain against the dollar would cut operating profits by approximately ¥34 billion ($424 million). Kyodo News conducted a survey of 105 large companies at the end of July. Fifty-five companies stated that they were aiming to relocate more of their divisions overseas in order to counter the effects of the strong yen, falling share prices and power shortages.

On 4 August the yen had appreciated to close to the 17 March all-time low of ¥76.25. At the request of MOF the BOJ intervened unilaterally in the foreign exchange market, selling approximately a record ¥4.66 trillion ($59.2 billion) to lower the exchange rate to near ¥80 per dollar. The yen's appreciation reflected the economic difficulties in the United States and the Eurozone and intervention in FOREX to prevent the yen's rise would ultimately prove futile. The downgrade of US government bonds from AAA to AA+ by S&P a few days later served to raise the yen to the ¥76 mark.

Simultaneously, the BOJ embarked on another bout of Quantitative Easing, injecting ¥10 trillion into the economy to include purchasing ¥5 trillion of assets, ¥2 trillion in JGB and ¥2 trillion in short-term government bills. The central bank also purchased commercial paper and broadened its loan program. Unfortunately, due to the uncertainty in international markets and coupled with the Nikkei tumble during the same week, the bank's two policy tools made little impact on the yen's exchange rate which rose to ¥75.95 on 19 August. There was speculation that due to the US Fed's intention to maintain interest rates at 0.0 percent until 2013 the yen would remain strong.

The foreign exchange special account is generally used by the government and the BOJ to intervene in FOREX. However, on 24 August, the MOF announced that to combat the appreciation of the yen the government would set up a one-year fund worth ¥7.6 trillion ($100 billion) to provide low-interest rate loans to companies to encourage corporations to swap yen reserves for dollar holdings and purchase equity in foreign companies. JBIC managed the program. Critics questioned the fund's effectiveness since it was considered inadequate. Additionally, many companies were already taking advantage of the strong currency and purchasing foreign companies. There were also worries that the program would serve to take production out of Japan.

Besides the domestic problems previously discussed, there were also problems in foreign markets to contend with, including the European debt crisis, the turmoil in the Middle East where Japan had significant investment, a conflict of interests with China over gas fields and the Senkaku Islands, and the conflict with South Korea over the Dokdo Islands. Also, higher commodity prices and the substantial increase in the importation of liquid gas and oil would put a heavy burden on consumers and the already beleaguered utility industry:

1. TEPCO imported annually 20 million metric tonnes of liquefied natural gas (LNG) or approximately 20 percent of Japan's total

imports. The nuclear crisis fueled the further rise in LNG prices by June which was about $14 for one million BTU. Demand from Russia, the largest importer, China, India and Japan had already pushed prices up on the international market and Japan's imports of the gas since 2009 had tripled the short-term price. Since TEPCO, KEPCO, CHUDEN, KYUDEN and YODEN, which primarily rely on their nuclear power plants to supply electricity, are substituting thermal power from their thermal power plants to compensate for the loss of nuclear power, according to the Japan Oil, Gas and Metals Corp. (JOGMEC), the short-term market price for LNG was expected to be $20 by autumn and further escalate by the winter months. JOGMEC, a METI IAI, was considering increasing government investment in natural gas exploration in 2012. On 23 July, IMPEX Corp., which engages in projects with JOGMEC (see p. 43), announced that it would sell a 30 percent stake in a floating liquefied natural gas plant off the coast of Indonesia to Shell Oil in return for technical know-how in the development of the plant.

2. Japan also anticipated that oil imports by the power utilities would triple. Japan, which was already the world's third largest importer of crude oil, would be importing an additional 350,000 barrels per day, which was selling for $100 per barrel in July. CHUDEN predicted that its costs would rise by ¥250 billion ($3.1 billion) due to the procurement of fuel for thermal power generation. Consumers' utility bills, already among the most expensive among the industrialized countries, would rise considerably by winter.
3. If all nuclear power reactors are suspended and thermal power is increased, it is predicted that carbon emissions will increase in 2012 by 15 to 16 percent from FY 1990 levels.

Moody's senior vice-president Thomas Byrne attributed the downgrade of Japan's debt in August to the frequent changes of prime ministers which were hampering the government from rebuilding finances, to power shortages and to concerns about the government's ability to implement policy.

For twenty years successive LDP and DPJ administrations struggled to plan and implement policies that would bring Japan out of deflation, cut the government deficit, support the growth of small and medium-size businesses to ignite regional economic development, and sort out the social security system. But the model of government administration was enmeshed in the vested interest of the players in Japan Inc. Indeed, the opposition from his cabinet forced Kan to reconsider his initial proposals to raise the consumption tax from the existing 5 percent to 10 percent by 2015 and implement fiscal discipline in order to halve the budget deficit by 2016 and balance the budget by 2021. The new target date set for an increase of the consumption tax was 2016.

Naoto Kan was Japan's sixteenth prime minister since 1989 but the DPJ together with the BOJ had followed the same fiscal and monetary policies as had the LDP. During the DPJ's reign the yen appreciated more than 23 percent against the dollar, consumer prices remained below the BOJ's inflation target of 0.1 percent and sovereign debt continued to rise. Kan's public approval rating before 11 March was 17 percent. By July, ratings slumped to 15 percent and, although 70 percent of voters applauded Kan's motives for ending a nuclear-dependent society, they wanted Kan to resign by August. The public opinion poll taken by *Kyodo News* on 1 August showed that the DPJ's support rate was 14.7 percent and the LDP's support was 25.9 percent. After only two years in power the DPJ was imploding and the LDP was making a comeback. The constant inter-party bickering and squabbling resulted in institutional paralysis. The Japanese system of government administration was in meltdown and the Japanese were falling into an economic and political morass.

Carpenter (2008) stated:

> Even if the Democratic Party of Japan were swept into the majority in both Houses by a rising tide of discontent among voters and even if structural reforms are initiated, the effort would be akin to plugging holes in a dyke to stem the flood of problems that are a consequence of years of reliance on an antiquated system and pork-barrel policies.[15]

Japan Inc. is alive and well

The main body of literature on Japan's political economy has simply defined elements in the governing system that were once considered to be the principal reasons for Japan's economic might. Emphasis was placed on the power of the bureaucracy and the domination of a single political party. *Amakudari* was defined as the migration of former bureaucrats to positions in the private sector where they forged an abiding relationship between government and business, hence the coining of the term 'Japan Incorporated'.

The development of Japan's electric power companies and the nuclear power industry through the close collaboration of businesses, ministry officials and elected officials and the intimate roles that Special Corporations, research institutes and industrial associations played during the process is a superb representation of Japan Inc. There was conjecture that Kan's political demise was due, in part, to his attempts to reform the model, in this case the nuclear power and electric power industries, which are primary to national policy as well as to the interests of politicians, bureaucrats and businessmen.

Although commentators have written articles that mentioned *amakudari* in public corporations, they explained how the migration process worked without analyzing the effects of the migration. Furthermore, little heed was paid to how public corporations could operate as extensions of the ministries and as organizations that came to promote the interests of elite officials. Since the 1990s successive administrations attempted to bust the bureaucracy's hold on the political economy by eliminating *amakudari*, downsizing Special Corporations and limiting the contact between bureaucrats and law-makers who were in decision-making roles.

Since the beginning of the second lost decade, the IMF and World Bank had been prodding the government to initiate structural reforms without considering that the very nature of the structure of Japan's government defied reform and that at the heart of government were the ministries' public corporations. When Koizumi left office in 2005, there were fifty-seven IAIs managed by the ministries. By 2010, the number had increased to 101.

Back to basics: a variation on the '1955' theme

Masahisa Saito told his audience at Georgetown University in 1994 that the Japanese people had relied for centuries on either an emperor or a military regime to govern them and did not want to take the initiative to plan their own destiny, preferring to entrust responsibility to a bureaucracy. The elements intrinsic to the post-war governing system were:

1. A rigid hierarchical socio-political system with a bureaucracy invested with the powers to plan and implement Japan's post-war economic growth without being subject to legal sanctions.
2. The close cooperation between business, the bureaucracy and the National Diet that began during the Meiji Restoration. The finely tuned relationship continued before and after the Second World War.
3. A conservative political party that supported the bureaucracy's policies consistently during Japan's post-war period.
4. The network of former bureaucrats and bureaucrats throughout business and government that is perpetuated by the *amakudari* system,

which greatly enhanced bureaucratic power to enforce policy at both the national and local government levels.
5. Significant pressure to accept bureaucratic policies and guidance.

The ministries began experiencing a brain-drain as young civil servants chose to leave the ministries mid-career when some of the ministries merged in 2001. Since the ministries had individual institutional cultures the mergers created an identity crisis among young officials who were concerned about future promotion. Also, the prospective post-retirement careers in industry were decreasing as bureaucrats' reputations were being tarnished by scandal. As importantly, territorial or 'turf' issues were taking precedence over policy planning. Morale was low.

The younger officials were also discouraged by the structural rigidity of their organizations. Officers who had been sent to foreign universities to receive degrees often returned to responsibilities that were unrelated to their acquired expertise. The insularity of the ministries pressured officers who had been 'loaned' to branches of IAIs located in a foreign country for periods of two to three years to forget what they experienced while living outside of Japan in order to integrate smoothly again into the workplace. Moreover, their new assignments did not take into consideration their previous duties.

During the Meiji Restoration, Japan's industrial development was driven by the collaboration between the bureaucracy and big businesses. During the war effort, the collaboration continued. When the Cold War began in 1945 SCAP recognized that in order to secure an industrially strong and conservative ally in the Pacific as swiftly as possible it was paramount to the United States' interests to rely on the wartime bureaucrats and their business partners to rebuild Japan's industrial complex. It was also essential to install a conservative political regime that backed the US–Japan Security Treaty.

The 1955 one-party system remained intact until 1990 and the majority of voters were isolated from the political process. They perceived their government officials as being separate from themselves. Although there have been reforms of the electoral system, there are few mechanisms in the Japanese socio-political system that promote voter participation in the political process. The network of interconnecting formal and informal relationships between politicians, bureaucrats and businessmen continues and this organic interdependence paralyzes the implementation of policy during crisis.

Besides the progeny of Nobusuke Kishi, Ichiro Hatoyama and Shigeru Yoshida who were discussed in Chapter 5, a number of key politicians who are in the LDP and the DPJ, including LDP president Sadakazu Tanigaki,

Prime Minister Noda and Seiji Maehara, share similar ultra-conservative ideologies and nationalistic sentiments and vigorously support the US–Japan Treaty. On 29 August, two days before he was elected DPJ president, at a news conference Noda alluded to China as 'a nation that is mixing up economic growth and nationalism'. A week later, as the newly elected Prime Minister, he vowed to strengthen Japan's defense capabilities to protect territorial rights. Noda also pledged on 11 September that his government would continue the efforts to bring all of the Japanese who were abducted by North Korea back to Japan. Seiji Maehara is very popular among young voters. His debut as DPJ Policy Chief was on 7 September at a symposium for Japanese and US security experts in Washington DC. Maehara's keynote speech, titled 'Japan–U.S. Alliance in Multilateral Cooperation', called for the review of the five principles that strictly limit the use of weapons by the SDF in UN peacekeeping operations. He emphasized the need to relax the restrictions in order to enable Japan to give full support to the United States and other foregin forces if they were to come under attack. He also sought a review of the three principles that formally banned the export of weapons and advocated the relaxation of regulations to enable Japan to develop and produce weapons with the United States and other countries. Maehara referred to China as a 'game changer' and a nation trying to revise international rules. Their views are also shared by many Japanese which indicates a trend towards a more insular society. The DPJ's overtures to the LDP to form a coalition government and the party's acquiescence to discard key pledges stated in their 2009 manifesto signal the return of a one-party system.

Before the 11 March crisis, voters were yearning for a strong leader to guide Japan's economy out of fifteen years of deflation and tackle the increasingly complex problems that were hampering economic growth. The rapid turnover of administrations since 2006 and the continuing deterioration of the economy have engendered a significant mistrust of politicians and an ever-present anxiety among voters. The Japanese are becoming risk-averse and inward-looking.

One of Prime Minister Noda's priorities is to take action to stem the yen's appreciation. Another is to raise taxes. Keidanren supports these measures and, along with Japan's largest labor federation, Rengo, will participate in Noda's new economic panel to form economic policies. Noda's intention to raise taxes is pragmatic fiscal policy and supported by Keidanren. His support for the restart of the nuclear reactors and the reliance on nuclear power for electricity, at least in the medium term, is also pragmatic because Japanese manufacturers will be persuaded to postpone relocating some operations overseas. Even though Noda has asked the opposition parties to join with the DPJ to form a unified government, there are misgivings about whether Noda's administration will have any more success than previous administrations in tackling Japan's profound economic problems. By 14 September, the LDP was demanding a snap election in the Lower House. Undoubtedly, unless the squabbling stops, political gridlock will continue to undermine the implementation of policies.

Kan pronounced the earthquake-tsunami disaster as Japan's worst crisis since the Second World War. There will be massive investment by the government for the reconstruction of Tohoku and for the support of small businesses and exporters in an attempt to keep production in Japan. There will be investment in public works projects and in the development of renewable energy resources. After the war ministry officials played a vital role in the planning and implementation of policies which served to resuscitate Japan's industrial complex and economy. Noda intends to re-establish an alliance with the ministries to help him and his administration through this time of crisis.

The crisis offers Prime Minister Noda the opportunity to beef-up the bureaucracy through the institutional restructuring of the ministries and to cultivate technocrats who will collaborate with politicians in the planning of strategy and cohesive policies.

The DPJ manifesto calls for the elimination of *amakudari* and the dismantling of the ministries' public corporations. However, before eliminating the system, there must be other mechanisms in place that act to incentivize talented university graduates to enter the civil service.

The author recommends that the ministries, which were merged in 2001, be returned to their former state to alleviate turf wars between the merged organizations and to resuscitate the individual corporate cultures. Furthermore, university graduates would welcome the civil service if certain conditions existed that supported a more egalitarian environment such as the discontinuation of the seniority system and the installation of a 'promotion by merit' system. Additionally, wages should represent corporate pay-scales, retirement benefits should be increased and substantial bonuses awarded for good work. Adequate administrative staff should also be provided. It is suggested that a special corporate tax levied on the large exporters would support the financing of the restructuring and bring in the best and the brightest.

The reorganization of the ministries would serve to assuage the hierarchical structure and discourage patronage and a 'yes-man' society. It would also serve as a natural process in the reform of an antiquated model of administration which is stalling the implementation of economic and social policies that can resolve the unremitting socio-economic problems that have been challenging the Japanese for years. Time is of the essence.

Notes

Introduction

1. The Great Kanto Earthquake followed by a tsunami in 1923 measured a 7.9 magnitude and killed 140,000 people.
2. The Nuclear Industry and Safety Agency (NISA), established in 2001 by the Ministry of Economy, Trade and Industry, regulates safety standards in the nuclear power sector. The Science and Technology Agency prior to 2001 had been involved in the regulation of the nuclear power sector.
3. A sievert is a measurement used by scientists to assess the biological effect radiation has on people and the possible harm. A millisievert is a thousandth of a sievert. The International Commission on Radiological Protection recommends the legal limit for radiation for the general public from nuclear power plants is 1 millisievert (mSv) a year. In emergency conditions the upper limit is 5 mSv. The current limit for employees in the nuclear industry is 20 mSv a year for a maximum of five years or 100 mSv for five years. The lowest level at which there is any long-term increase in cancer risk is 100 mSv a year.
4. The Japan Nuclear Safety Organization (JNES) is an Independent Administrative Institution managed by METI.
5. Also known as radioiodine, iodine-131 is a radioactive isotope. As a major plutonium and uranium product it can cause the mutation and death of cells it invades. It is also used in medications.
6. TBq is a unit of radiation.
7. Bloomberg.com (16 June 2011).
8. 'Ex-advisor slams Kan, NSC for local's exposure' (*Japan Times*, 11 June 2011).
9. METI's Agency of Natural Resources and Energy (ANRE) was established in 1973 to develop energy sources. The agency oversees the development of energy sources, including nuclear.
10. Kyodo News Service, 8 May 2011.
11. The Ministry of Economy, Trade and Industry's administrative jurisdiction encompasses the manufacturing, small business, retail, and energy sectors. METI plans industrial and economic policies, sets industrial standards, issues licenses and patents and plays a significant role in representing Japan at trade negotiations. NISA has over 800 staff working in 16 divisions, among them Nuclear Power Licensing, Nuclear Safety Regulatory Standards, Nuclear Power Inspection, Policy Planning and Coordination, Nuclear Emergency Preparation, Industrial Safety Inspection, Electric Power Safety and Electricity and Gas Industry.
12. S. Carpenter, *Special Corporations and the Bureaucracy: Why Japan Can't Reform* (Basingstoke: Palgrave Macmillan), p. 121.
13. Ibid.
14. S. Carpenter, *Why Japan Can't Reform: Inside the System* (Basingstoke: Palgrave Macmillan, 2008), p. 143.

15. Hatoyama, Japan's first DPJ prime minister resigned from office because he failed to fulfill his promise to the electorate to resolve the contentious issue of the relocation of the US Marine Corps Air Station Futenma to another area on Okinawa. With 47,000 American troops, 75 percent of American military installations in Japan are located on 10 percent of Okinawa land because of its close proximity to Southeast Asia, namely China and Taiwan. Both Okinawa residents and mainland Japanese have protested for years about the US military presence. In 2003, Japan and the US agreed to relocate Futenma to another area on Okinawa by 2008. In 2006, the popular resort town of Nago on northern Okinawa was chosen to host the base. Nago residents led by the mayor protested. However, due to political dissension the Japanese government postponed the decision to move. The US demanded that a decision be reached by May 2010 but Hatoyama continued to procrastinate because Nago's newly elected mayor opposed the move and Okinawa's governor Hirokazu Nakaima, who was pressuring Hatoyama to abide by the 2006 agreement, wanted changes in the agreement regulating the status of status of US troops in Japan After his resignation Hatoyama admitted that he had not considered strategies for the relocation before he made his promise to the Japanese to deal with the sticky situation.

1 Independent Administrative Institutions: In Name Only

1. C. Johnson, *Japan: Who Governs? The Rise of the Developmental State* (New York: W. W. Norton, 1995), p. 129.
2. Ibid.
3. The ALIC was converted to an IAI on 1 October 2003. It is still managed by the Ministry of Agriculture, Forest and Fisheries.
4. The abduction and rape of a twelve-year-old young girl in Futenma by three servicemen stationed at the US Marine Corps Air Station Futenma exacerbated the opposition to the US–Japan Security Treaty. The incident triggered the American–Japanese decision in 1995 to move the base to another location.
5. *Mainichi Shimbun*, 1994.
6. The ministries' headquarters are located in Kasumigaseki, a district in Tokyo.
7. Japan Government Bonds (JGB) include deficit-covering bonds, bonds for construction and FILP bonds. FILP bonds are issued on the market as JGBs with the same market price. Interest rates are the same.
8. K. Ishii, *Bureaucrat Heaven: The Bankrupting of Japan* [*Kanryo Tenkoku Nihon Hassan*] (Tokyo: Michi Shuppansha, 1999).
9. K. Ishii, *The Parasites Gobbling Up Japan: Dismantle All Special Corporations and Public Corporations!* [*Nihon wo Kuitsuku Kiseichu: Tokushuhoujin Koeki Houjin wo Zenhai Seiyo!*] (Tokyo: Michi Shuppansha, 2001).
10. Nobuteru is the son of Shintaru Ishihara, a popular nationalist politician who was elected to a fourth term as governor of Tokyo in April 2011.
11. K. Tsutsumi, *The Monster Ministries and Amakudari: White Paper on Corruption* [*Kyodai Shocho Amakudari Fuhai Hakusho*] (Tokyo: Kodansha, 2000), p. 190.
12. R. Yoshida, 'Robbing of Pork Barrel had LDP Squealing', *The Japan Times* (16 June 2001).

13. www.jhf.go.jp
14. Asset liability management (ALM) is a tool used to manage risks such as liquidity risks, interest rates risk, credit risk and operational risk.
15. www.ur-net.go.jp
16. www.jogmec.go.jp
17. Tsutsumi, *The Monster Ministries and Amakudari*, pp. 145–75.
18. The BSCs are now known as IBSCs or International Business Support Centers.
19. E. Lincoln, *Troubled Times: US–Japan Trade Relations in the 1990s* (Washington, DC: Brookings Institution, 1999), p. 102.
20. Holstein was a former world editor of *Business Week* (1984–94).
21. W. J. Holstein, 'With Friends Like These', *U.S. News and World Report* (16 June 1997), p. 48.
22. Ibid.
23. Ibid., p. 47.
24. www.nedo.go.jp
25. www.jnes.go.jp
26. www.aist.go.jp
27. S. Carpenter, *Special Corporations and the Bureaucracy: Why Japan Can't Reform* (Basingstoke: Palgrave Macmillan, 2003), p. 123.

2 *Amakudari* in the Ministries' IAIs, Public Corporations, Research Institutes and Affiliated Agencies

1. E. Lincoln, *Troubled Times: US–Japan Trade Relations in the 1990s* (Washington, DC: Brookings Institution, 1999), p. 190.
2. K. Calder, 'Elites in an Equalizing Role: Ex-Bureaucrats as Coordinators and Intermediaries in the Japanese Government Relationship', *Comparative Politics*, 21, 4 (1989): 379–403 (p. 389).
3. When the author was working in the Tokyo headquarters of a large Japanese liquor and pharmaceutical producer in 1988 all of the computers used by the corporation were produced by the giant Nippon Electric Corporation (NEC). The following year when the author returned to the office from a short holiday, she found that all of the computers in every division had been replaced by IBM computers. According to colleagues, the director of Human Resources had made the decision to purchase IBM equipment without consulting the president.
4. See Introduction.
5. K. Tsutsumi, *The Monster Ministries and Amakudari: White Paper on Corruption* [*Kyodai Shocho Amakudari: Fuhai Hakusho*] (Tokyo: Kodansha, 2000), pp. 145–75.
6. *The Japan Times Online* (3 May 2011).
7. Ibid.
8. The President of JETRO New York (1994–6) is an example of how METI officials can be transferred towards the end of their career. The official was 'loaned' to JETRO headquarters in Tokyo where he served as the Director of the Planning Division. He returned to the ministry for two years before being 'loaned' to JETRO San Francisco where he served as the president for three years. He was sent directly to JETRO New York where he was president for two years. He then migrated to IBM Japan.

9. An example of the consequences of interpersonal networks established between METI officials and industries in their ministry's administrative jurisdiction is documented (Carpenter 2008) by an incident that occurred in 2005 involving a METI official who used funds from the Japan Keiren accounts held by the Secretariat Planning Office.

   The Industrial Revitalization Corporation of Japan (IRCJ) was established by government on 16 April 2003 as a fully-owned subsidiary of the Deposit Insurance Corporation of Japan and entrusted with $80 billion to address the issue of non-performing loans. Corporate expansion in Japan during the bubble economy years of 1986–90 was funded by substantial loans from national and regional banks. Many of the loans were underwritten by the rising value of real estate. After the bubble burst and real estate values fell sharply, the banks were left with a large body of non-performing loans (NPL) or 'bubble loans'. NPL are loans that are not currently accruing interest or on which interest is not being paid.

   The official function of the IRCJ was to purchase loans from creditors and remove conflict of interests between the main bank and other creditors. In this way, the rehabilitation of the borrowers could be facilitated. The IRCJ selected failing firms that were considered suitable for rehabilitation, acquired the liabilities of the firms, persuaded the banks to forgive a substantial amount of the debt, thus eliminating NPL, selected a firm or firms deemed appropriate for sponsoring capital investment in the floundering company and, together with the sponsoring firm, implanted a restructuring program. The IRCJ was given two years (until 31 March 2005) to accept applications from banks and their major clients and to approve assistance of the firms, and three subsequent years to dispose of all liabilities.

   The IRCJ was not tied to any one ministry, but the firms in the sectors being assisted by the IRCJ represented construction and transportation companies in the administrative jurisdiction of the MILT and manufacturing and retailing (METI). A major firm chosen for restructuring was Kanebo Ltd., a textile, food and cosmetic conglomerate. When Kao, Japan's leading manufacturer of toiletries and household and industrial cleaners, offered to take over the cosmetic division for $4.9 billion in 2004 Kanebo was saddled with debts of $6 billion. But Kanebo refused the offer and asked its creditors to forgive $1.2 billion in loans. However, the banks refused and the conglomerate sought the help of the IRCJ, who purchased its core cosmetic division for $3.2 billion.

   Kanebo's decision was not entirely in its best interests. In May 2005, the IRCJ charged the popular company with accounting fraud. Evidently, Kanebo had a negative worth for the previous nine years, but it had disguised its financial troubles by altering the books. Furthermore, Price Waterhouse Coopers Aoyama, Kanebo's auditors, had cooperated in the effort. On 13 June 2005, the Tokyo Stock Exchange (TSE) de-listed Kanebo, whose stocks had been publicly traded for 114 years. PWC Aoyama was sanctioned and suspended from auditing for two months.

   The investigation into the accounting fraud also confirmed the interpersonal network between METI and the IRCJ. In June, Taizo Nakatomi, a high-ranking METI official and former head of METI's policy planning office, admitted that in 2004 he withdrew ¥29 million ($270,000) from the bank accounts where funds for the IRPI research projects were deposited.

In addition to the $335,000 he used his personal funds to purchase Kanebo shares in 2004. However, he had since returned the money plus $24,000. METI Vice-Minister Hideji Sugiyama denied that there was insider trading and asserted that Nakatomi did not have access to special information. Although Nakatomi resigned from METI, he was not arrested for embezzlement.

10. The failure in 1996–8 of three major financial institutions, followed in 1998 by the failure of Nippon Credit Bank (NCB) and the Long-Term Credit Bank (LTCB), both Special Corporations that had been established by MOF, revealed the severity of NPL. Until the early 1990s and the establishment of a Financial Services Agency (FSA) banks regarded both MOF and the BOJ regulators as their protectors who shielded them from close scrutiny by outside auditors. The FSA initially was considered independent from the MOF and the BOJ and the first director was hired from outside both agencies. However, subsequent directors have been MOF officials.

    In 1997, a scandal surrounding the entertainment of MOF and BOJ officials at expensive restaurants and clubs by banks brought to public attention that the non-disclosure of NPL by the banks was partially the consequence of the intimate relationship between the BOJ, MOF and the banks. In their defense, the banks insisted that they had been pressured by the officials to provide the entertainment. Many retired bureaucrats from MOF and retired executives are employed in top management positions in metropolitan banks and financial institutions.
11. Bloomberg reported on 29 April 2011 that Orix Corp., Japan's third-largest corporate borrower in 2010, was renewing its unused $600 million credit line from overseas lenders which included BNP Paribas SA and Crédit Agricole SA. Orix was one of the first Japanese companies to issue bonds when its sales declined after the 11 March earthquake.
12. On 15 September 2004, Murakami met with Horie and recommended shares in Nippon Broadcasting System (NBS) as a very good buy. On 1 October, M&A purchased 18 percent of the radio network system. On 8 November, Murakami met again with Horie, who discussed plans to buy 35 percent of NBS, this enabling Livedoor to join forces with M&A to take over the company. Murakami proceeded to buy 1.9 million additional shares in NBS from 9 November 2004 to 26 January 2005 for ¥9.95 ($90 million) and then sold approximately 5 million shares in February when Livedoor bought a large number of shares in a takeover bid for NBS, which was initially foiled by Fuji Television, a company that was connected to NBS through cross-shareholdings. The Murakami Fund reaped profits of ¥3 billion ($27 million) while Murakami personally earned ¥150 million ($1.35 million) from the transaction.
13. In April 2011, it was estimated that the population of children under the age of 15 had fallen 90,000 from the previous year for the thirtieth consecutive year. At 16.93 million, the number marked a record-low since 1950.
14. In 1997, the SIA began consolidating three different databases, causing significant problems in data collection.
15. www.nenkin.go.jp
16. Many Japanese prefer putting their savings into postal savings because the government offers slightly higher interest rates. They also feel more secure because the government manages the institution.
17. See note 7.

18. Tokyo governor Shintaru Ishihara, a right-wing nationalist, named the party.
19. R. Perry, *The Times*, 31 August 2009.
20. S. Carpenter, *Why Japan Can't Reform: Inside the System* (Basingstoke: Palgrave Macmillan, 2008), p. 147.

3 'Information-Sharing' is Not a Buzz-Word in Japan

1. A. Mikuni and R. Murphy, *Japan's Policy Trap: Dollars, Deflation, and the Crisis of Japanese Finance* (Washington, DC: Brookings Institution, 2002), p. 145.
2. S. Carpenter, *Why Japan Can't Reform: Inside the System* (Basingstoke, UK: Palgrave Macmillan, 2008), p. 35.
3. www.pressnet,or.jp/eng/
4. Ibid.
5. A. Gamble and T. Watanabe, *A Public Betrayed: An Inside Look at Japanese Media Atrocities and their Warnings to the West* (Washington, DC: Regnery Publishing, 2004).
6. T. Iwase, *The Reason Newspapers are Not Interesting* [*Shimbun ga Omoshiroku nai Ryuu*] (Tokyo: Kondansha, 1998).
7. Shmel Gvido, 'Kisha Clubs', *Japan Media Review* (www.orj.or/japan/wiki/Kishaclubs/, 2006).
8. L. Freeman, 'Japan's Press Clubs as Information Cartels', JPRI Working Paper No. 18, Japan Policy Research Institute (San Francisco, CA, April 1996). Laurie Ann Freeman also published *Closing Shop: Information Cartels and Japan's Mass Media* (Princeton: Princeton University Press, 2000). The publisher states that the book describes how the network of interrelationships between the media industry and corporate groups can control the reporting of political news to the public and protect the political and corporate elite from close scrutiny.
9. 'PM yet to keep promise of open press conferences', *Japan Today* (courtesy of Kyodo News, www.japantoday.com, 8 January 2010).
10. J. Hongo, 'Group seeks "kisha club" shutdown', *The Japan Times Online* (www.japantimes.co.jp, 20 April 2010).
11. K. Takahara, 'Tight-lipped TEPCO lays bare exclusivity of press clubs', *The Japan Times Online* (3 May 2011).
12. 'DPJ crisis panel exec found playing golf', *The Japan Times Online* (Kyodo News), www.japantimes.co.jp (6 May 2011).
13. China overtook Japan as the second largest economy in 2010.
14. Carpenter, *Why Japan Can't Reform*, p. 36.

4 Elements Intrinsic to Japan's Political Economy

1. K. Calder, 'Domestic Constraints and Japan's Foreign Economic Policy of the 1990s', *International Journal*, 46 (Autumn 1991), p. 612. C. Johnson, *MITI and the Japanese Miracle: The Growth of Industrial Policy, 1925–1975* (Stanford, CA: Stanford University Press, 1982).
2. C. Johnson, *Japan: Who Governs? The Rise of the Developmental State* (New York: W. W. Norton, 1995).

3. www.kantei.go.jp/foreign/constitution_and_government_of_japan/fundamental
4. A. Mikuni and R. Murphy, *Japan's Policy Trap: Dollars, Deflation, and the Crisis of Japanese Finance* (Washington, DC: Brookings Institution, 2002), pp. 40–1.
5. Japan's ministry officials referred to MITI as 'notorious MITI' because of the ministry's propensity to pursue the territory of other ministries.
6. See Chapter 1.
7. Referred to in Chapter 2.
8. Yasufumi was elected on the LDP ticket to the Lower House for the first time in 1997. He won two more terms in 2000 and 2002. He served in Koizumi's 2004 and 2005 cabinets as minister of state for science and technology policy, minister of state for food safety and minister of state for information technology.
9. *Mainichi Shimbun* Editorial Staff, *The Kasumigaseki Syndrome* [*Kasumigaseki Shindoromu*] (Tokyo: Mainichi Shimbunsha, 1994), p. 46.
10. Japanese businesses and government agencies are among the clients of the law firm Marks & Murase. It is interesting to note that Carl J. Green, who was appointed the director of ALPS at the time of Naito's professorship, was appointed the senior representative of the Hitachi Corporation in Washington, DC in February 1997.
11. The agency was integrated into the Ministry of Land, Infrastructure, Transport and Tourism in 2000.
12. Hence the name 'notorious MITI'. Chalmers Johnson gives a good example of this behavior in 'MITI, MPT, and the Telecom Wars How Japan Makes Policy for High Technology', in C. Johnson, L. Tyson and J. Zysman (eds), *Politics and Productivity: The Real Story of Why Japan Works* (Cambridge, MA: Ballinger, 1989), pp. 177–239.
13. *Mainichi Shimbun* Editorial Staff, *The Kasumigaseki Syndrome*, pp. 296–308.
14. J. Sager, The Origins of Japan's Economic Philosophy', *JPRI Critique*, 7, 9, Japan Policy Research Institute (San Francisco, CA, October 2000).
15. Ibid.
16. Ibid.
17. From 1895–1900 France, Germany, England and Russia made large outward investments in China through mining and railroad concessions and through the leasing of territories in those specific areas. The United States annexed Hawaii and the Philippines in 1898.
18. The Taiwanese learned Japanese in school as a second language.
19. Recession cartels are formed to rationalize production so that major companies affected can continue to operate on a level playing field.
20. Emperor Showa, who was known more commonly as Emperor Hirohito, reigned until 1989, the longest reign for a Japanese ruler.
21. The Council's website explains that the 'official organization responds to enquiries from the Ministry of Economy, Trade and Industry on important topics related to METI's policy, particularly improving the economic strength of the private sector and promotion of good international economic relations'.
22. In his book *MITI and the Japanese Miracle* Chalmers Johnson defined industrial rationalization as a 'state policy at the micro-level, state intrusion into the detailed operations of individual enterprises with measures intended to improve those operations (or, on occasion, to abolish the enterprises)', p. 27.

23. Ibid., p. 104.
24. Ibid., pp. 109–10.
25. From 1922 to the end of the Second World War Japan was served by twenty-two prime ministers and eleven were either retired generals or admirals with the exception of Tojo, who was an acting general.
26. 'Manchuko: Distress Goods', *Time Magazine* (25 March 1935), www.time.com
27. The IBJ remained an important financer of industry during Japan's post-war economic growth until 2001, when, due to non-performing loan problems, it was merged with two private metropolitan banks.
28. T. Nakamura, *The Postwar Japanese Economy: Its Development and Structure, 1937–1994* (Tokyo: University of Tokyo Press, 1977), p. 18.
29. Ibid., p. 40.
30. The constitution is often referred to as the 'Pacifist Constitution'.
31. Johnson, *MITI and the Japanese Miracle*, p. 149.
32. R. A. Werner, *Princes of the Yen: Japan's Central Bankers and the Transformation of the Economy* (Armonk, NY: M. E. Sharpe, 2003), p. 131.
33. Joseph Dodge, chairman of the Detroit Bank, was SCAPs economic advisor who planned policies to stabilize Japan's economy.
34. Werner, *Princes of the Yen*, p. 25.
35. K. Yamamura, 'Success that Soured: Administrative Guidance and Cartels in Japan', in K. Yamamura (ed.), *Policy and Trade Issues of the Japanese Economy: American and Japanese Perspectives* (Washington, DC: University of Washington Press, 1982), pp. 77–112 (p. 99).
36. R. Katz, *Japan: The System That Soured* (New York: M. E. Sharpe, 1998), p. 45.
37. A. Mikuni, 'Why Japan Can't Reform Its Economy', JPRI Working Paper No. 44, Japan Policy Research Institute (San Francisco, CA, April 1998).
38. Mikuni and Murphy, *Japan's Policy Trap*, p. 96.

5 The DNA of Japan's Post-war Political System

1. R. Samuels, R., 'Kishi and Corruption: An Anatomy of the 1955 System', JPRI Working Paper No. 83, Japan Policy Research Institute (San Francisco, CA, December 2001).
2. Ibid.
3. Ibid.
4. T. Umemura, www.asahi.com (20 July 2006).
5. C. Johnson, *Japan: Who Governs? The Rise of the Developmental State* (New York: W. W. Norton, 1995), p. 187.
6. 30 percent of law-makers in the Lower House have inherited their seats from their fathers, grandfathers and even from their great-grandfathers.
7. Kajima Corp. was one of the construction companies that built Fukushima Daiichi Power Plant.
8. www.mskj.or.jp
9. 'Q&A: Japanese Foreign Minister Seiji Maehara on China, the Yen', *Wall Street Journal* (25 September 2010).
10. Yukio Mishima (1925–70) was nominated for the Nobel Prize for Literature three times. He is considered one of Japan's greatest writers of the twentieth century because of his vast vocabulary and flawless use of Chinese characters.

He was born into a family of bureaucrats and was a nationalist. He enlisted in the Ground Defense Force in 1962. In the same year, Mishima formed a private army of which most of the members were students and pledged to protect the emperor, whom he felt exemplified the Japanese spirit. On 25 November 1970 Mishima, accompanied by four members of his army, tried to instigate a *coup d'état* at the SDF headquarters in Tokyo. They broke into the commandant's room and held the commander hostage. When Mishima read his manifesto to the soldiers congregated below, he was jeered. Humiliated, Mishima returned to the room and committed suicide.

11. D. McNeill, 'Japan must develop nuclear weapons, warns Tokyo Governor', Independent.co.uk (8 March 2011).
12. E. Johnston, 'Hashimoto steps up anthem fight', *Japan Times* (26 May 2011).

6 Pork-barrel Patronage in the Prefectures

1. www.tepco.co.jp
2. www.kepco.co.jp
3. www.chuden.co.jp
4. www.fepc.or.jp
5. *Asahi Shimbun* (25 June 2011).
6. jnfl.co.jp
7. Spent nuclear fuel contains plutonium, which can be extracted and used in new fuel but reprocessing is very expensive. Also, spent fuel emits lethal radiation, whereas separated plutonium can be handled easily. Consequently, the plutonium can be used to construct an atomic bomb. Anti-nuclear activists have demonstrated against the operation of the Rokkasho reprocessing plant.
8. After resigning from METI in 2006, Kodaira became an advisor in the same year for Mitsui Sumitomo Insurance Co. Ltd. He migrated to Toyota Motor Corporation to serve as a managing executive officer until June 2010 when he became the company's senior managing director, chief director of business development and chief director of information business.
9. Kusaka became the senior managing executive office at Mitsubishi Electric Co. in October 2010.
10. *Kyodo News*, 5 and 8 July 2004.
11. www.tepco.co.jp
12 Ibid.
13. 'METI OKs project for Nuclear Plant in Aomori', *Kyodo News* (31 December 2010).
14. C. Johnson, *Japan: Who Governs? The Rise of the Developmental State* (New York: W. W. Norton, 1995), p. 190.
15. Ibid., p. 193.
16. See Introduction.
17. www.iaea.org
18. K. Ishibashi, 'Why Worry? Japan's Nuclear Plants at Grave Risks from Quake Damage' (*International Herald Tribune/Asashi Shimbun*, 11 August 2007).
19. M. Fackler and N. Onishi, 'In Japan, a Culture That Promotes Nuclear Dependency' (*New York Times*, 31 May 2011).

20. Ochi also served as minister of transportation in Prime Minister Miyazawa's cabinet (1992–3) and as minister of agriculture in Prime Minister Hashimoto's 1997 cabinet.
21. Assessment based on government statistics.
22. 'The chrysanthemum revealed: probing the Japanese market' (*Entrepreneur*, Jan.–Feb. 1993).
23. *Kyodo News* (23 November 2006).
24 www.yoden.co.jp
25. '3 nuclear plants seek restart despite concerns', www.asahi.com (30 April 2011).
26. Standard & Poor's Ratings Services downgraded TEPCO's long-term corporate credit and debt rating from BBB+ to BBB in May. In June Moody's Investor Service cut TEPCO's bond ratings to junk status.
27. *Yomiuri Shimbun* (16 June 2011).
28. Ibid.
29. TEPCO and its creditor banks were opposed to the government's contention that there should be no limits to the power company's liability but accepted the compensation package on 12 May.
30. www.kyuden.co.jp
31. www.iaea.org
32. *Mainichi Shimbun* (22 June 2011).
33. *Kyodo News* (12 July 2011).

7 Japan's Nuclear Crisis: the Routes to Responsibility

1. www.japc.co.jp
2. www.jaea.go.jp
3. International Atomic Energy Agency, *Nuclear Energy Handbook* (www.iaea.org).
4. www.jaea.go.jp.jnc.jncweb
5. www.nsc.go.jp
6. World Nuclear Association, 'Tokaimura Criticality Accident' (Washington, 7 July 2007).
7. *Kyodo News* (29 April 2011).
8. *Nikkei* (15 July 2011).
9. '112 DPJ members in Lower House backed by TEPCO union' (*Japan Press Weekly*, 5 June 2011).
10. 'Hidden Agenda' (*Asahi News*, 16 June 2011).
11. 'Corporate Ethics and Compliance', www.tepco.co.jp
12. www.meti.nisa.gov.jp
13. *Yomiuri Shimbun* (4 July 2011).
14. www.jnes.go.jp
15. www.rwmc.org.jp
16. www.numo.or.jp
17. www.jaif.or.jp
18. www.jpower.co.jp
19. denken.or.jp
20. www.jnfl.co.jp
21. *Kyodo News* (22 June 2011).

8 The Japan System: Indestructible but Destructive

1. D. Ibison, 'FSA declares Japan's bad-loan crisis over' (*Financial Times*, 26 May 2005).
2. J. Choy, 'Japan's Non-Performing Loan Problem Refuses to Go Away' (Japan Economic Institute, 2000).
3. E. Lincoln, 'Making Some Sense of the Japanese Economy', JPRI Working Paper No. 94, Japan Policy Research Institute (San Francisco, CA, September 2003).
4. K. Kobayashi, 'The 15-Year War on Non-Performing Loans and Deflation', Research Institute of Economy, Trade and Industry (Tokyo, 2004).
5. M. Aoki, H. Patrick and P. Sheard, 'The Japanese Main Bank System: An Introductory Overview', in M. Aoki and H. Patrick (eds), *The Japanese Main Bank System: Its Relevance for Developing and Transforming Economics* (Oxford: Oxford University Press, 1994), pp. 1–50 (p. 32).
6. The Japanese overnight call rate is the interest that is charged when banks which are a part of the banking system loan to each other. The loans mature in one day and must be repaid by 10 a.m. the following day. BOJ uses the non-collateralized overnight base rates to influence the money supply. The rate affects the interest banks charge for commercial loans and for mortgages.
7. 'Japan Inc. turns a profit despite disaster', *Asahi News* (26 June 2011).
8. 'Honda and Toyota getting outgunned in the US' (*Japan Times*, 15 June 2011).
9. 'Bank Loans Fall in June for 20th Straight Drop' (Nikkei.com).
10. Daiwa Securities Group Inc. has similar operations.
11. *Kyodo News*, 19 June 2011.
12. Standard & Poor's, 27 January (www.standardandandpoors.com/ratings).
13. The highest Moody's long-term obligation rating is Aaa, which is considered the smallest degree of risk for investment. Aa (Aa1, Aa2, Aa3) bonds are regarded as 'very low credit risk' but the long-term investment risks can be greater.
14. 'Industrial Structure Vision: The Way Forward for Japanese Industry' (www.meti.go.jp/policy/economy/industrial/html).
15. S. Carpenter, *Why Japan Can't Reform: Inside the System* (Basingstoke: Palgrave Macmillan, 2008), p. 147.

Select Bibliography

Aoki, M., Patrick, H. and Sheard, P., 'The Japanese Main Bank System: An Introductory Overview', in M. Aoki and H. Patrick (eds), *The Japanese Main Bank System: Its Relevance for Developing and Transforming Economics* (Oxford: Oxford University Press, 1994), pp. 1–50.

Calder, K., 'Elites in an Equalizing Role: Ex-Bureaucrats as Coordinators and Intermediaries in the Japanese Government Relationship', *Comparative Politics*, 21, 4 (1989): 379–403.

Carpenter, S., *Special Corporations and the Bureaucracy: Why Japan Can't Reform* (Basingstoke: Palgrave Macmillan, 2003).

Carpenter S., *Why Japan Can't Reform: Inside the System* (Basingstoke: Palgrave Macmillan, 2008).

Choy, J., 'Japan's Non-Performing Loan Problem Refuses to Go Away', Japan Economics Institute (2000).

Freeman, L., 'Japan's Press Clubs as Information Cartels', JPRI Working Paper No. 18, Japan Policy Research Institute (San Francisco, CA, April 1996).

Freeman, L., *Closing Shop: Information Cartels and Japan's Mass Media* (Princeton: Princeton University Press, 2000).

Gamble, A. and Watanabe, T., *A Public Betrayed: An Inside Look at Japanese Media Atrocities and their Warnings to the West* (Washington, DC: Regnery Publishing, 2004).

Iishi, K., *Bureaucrat Heaven: The Bankrupting of Japan* [*Kanryo Tenkoku Nihon Hassan*] (Tokyo: Michi Shuppansha, 1999).

Iishi, K., *The Parasites Gobbling Up Japan: Dismantle All Special Corporations and Public Corporations!* [*Nihon wo Kuisuku Kiseichu: Tokushu Houjin wo Zenhai Seiyo!*] (Tokyo: Michi Shuppansha, 2001).

Johnson, C., *MITI and the Japanese Miracle: The Growth of Industrial Policy, 1925–1975* (Stanford, CA: Stanford University Press, 1982).

Johnson, C., *Japan: Who Governs? The Rise of the Developmental State* (New York: W. W. Norton, 1995).

Katz, R., *Japan: The System That Soured* (New York: M. E. Sharpe, 1998).

Kobayshi, K., 'The 15-Year War on Non-Performing Loans and Deflation', Research Institute of Economy, Trade and Industry (Tokyo, 2004).

Lincoln, E., *Troubled Times: US–Japan Trade Relations in the 1990s* (Washington, DC: Brookings Institution, 1999).

Lincoln, E., 'Making Some Sense of the Japanese Economy', JPRI Working Paper No. 94, Japan Policy Research Institute (San Francisco, CA, September 2003).

Lincoln, E. 'On Japan: Chilly Welcome', Council on Foreign Relations (New York, 26 February 2005).

Mainichi Shimbun editorial staff, *The Kasumigaseki Syndrome* [*Kasumigaseki Shindoromu*] (Tokyo: Mainichi Shimbunsha, 1994).

Mikuni, A. 'Why Japan Can't Reform Its Economy', JPRI Working Paper No. 44, Japan Policy Research Institute (San Francisco, CA, April 1998).

Mikuni, A. and Murphy, R. T., *Japan's Policy Trap: Dollars, Deflation, and the Crisis of Japanese Finance* (Washington, DC: Brookings Institution, 2002).

Nakamura, T., *The Postwar Japanese Economy: Its Development and Structure, 1937–1994* (Tokyo: University of Tokyo Press, 1977).

Okazaki, T., 'The Role of Holding Companies in Pre-War Japanese Economic Development', *Social Science Japan Journal*, 4, 2 (2001): 243–68.

Sager, J., 'The Origins of Japan's Economic Philosophy', *JPRI Critique*, 7, 9, Japan Policy Research Institute (San Francisco, CA, October 2000).

Samuels, R., 'Kishi and Corruption: An Anatomy of the 1955 System', JPRI Working Paper No. 83, Japan Policy Research Institute (San Francisco, CA, December 2001).

Tsutsumi, K., *The Monster Ministries and Amakudari: White Paper on Corruption* [*Kyodai Shocho Amakudari: Fuhai Hakusho*] (Tokyo: Kodansha, 2000).

Werner, R. A., *Princes of the Yen: Japan's Central Bankers and the Transformation of the Economy* (Armonk, NY: M. E. Sharpe, 2003).

Yamamura, K., 'Success that Soured: Administrative Guidance and Cartels in Japan', in K. Yamamura (ed.), *Policy and Trade Issues of the Japanese Economy: American and Japanese Perspectives* (Washington, DC: University of Washington Press, 1982), pp. 77–112.

Index

CPSIA information can be obtained at www.ICGtesting.com
Printed in the USA
LVOW072119131112

307219LV00006B/60/P

9 780230 354920